Und

Radioactive

Waste

N

Fifth Edition

Raymond L. Murray, Professor Emeritus
Nuclear Engineering Department
North Carolina State University
Raleigh, North Carolina

Edited by
Kristin L. Manke
Battelle Pacific Northwest Laboratories

 Battelle Press
Columbus • Richland

The first edition of this book was prepared by Pacific Northwest National Laboratory for the U.S. Department of Energy. Pacific Northwest National Laboratory is operated by Battelle.

Cover art courtesy of U.S. Department of Energy.
Cover design by Amy Madden and Cameron Bates.

Library of Congress Cataloging-in-Publication Data

Murray, Raymond LeRoy, 1920-
    Understanding radioactive waste / Raymond L. Murray ;
    edited by Kristin L. Manke.— 5th ed.
    p. c.m
    Includes bibliographical references and index.
    ISBN 1-57477-135-3 (alk. paper)
    1. Radioactive wastes. 2. Radioactive waste disposal.
I. Manke, Kristin L., 1967- II. Title
TD898.M87 2003
363.72'89—dc21   2003048110

Printed in the United States of America

Additional copies may be ordered through Battelle Press. Quantity discounts are available.

Battelle Press
505 King Avenue
Columbus, Ohio, 43201
Phone: 1-800-451-3543 or 614-424-6393
Fax: 614-424-3819
E-mail: press@battelle.org
Web bookstore: www.battelle.org/bookstore

To seekers of sound solutions

Acknowledgments

This book seeks to present facts about all aspects of radioactive wastes in a simple, clear, and unbiased manner. The information is intended for students and other interested or concerned members of the public.

The first edition of this book was prepared in 1982 by Pacific Northwest National Laboratory, operated by Battelle. Through subsequent editions, and now this new fifth edition, the book has benefited from the advice and assistance of many people, and we are appreciative.

Special thanks are due John V. Robinson for his conception of the book and continued advice and friendship over the years. Appreciation is extended to Judith Powell for her most capable editing of all previous editions.

Others who have provided moral support or valuable suggestions are Thomas D. Chikalla, L. Donald Williams, M. Ray Seitz, Edwin L. Fankhauser, Donald Fankhauser, Scott Nealey, Douglas Robinson, Margaret N. Maxey, David W. Levy, Frank K. Pittman, Eugene A. Eschbach, Lawrence J. Smith, Stanley M. Nealey, John T. Suchy, Harry Babad, Kenneth J. Schneider, Orville F. Hill, Harold H. Hollis, Dennis K. Kried, Elizabeth Reid Murray, Bonnie Berk, Thomas Overcast, Janie Shaheen, Richard Williamson, Faith Brennemon, Melvin Shupe, Paul Dunigan, Jack L. McElroy, Harry C. Burkholder, Ted Wolff, Steven C. Slate, and Walt Laity.

Illustrations have been prepared by Van Webb, Ann Kennedy, Valerie Stott, Luann Salzillo, Nancy Reid Baker, and Amy Madden. Thanks are due Sandra K. Blakley, Janet K. Tarantino, Donna Gleich, Darlene E. Whyte, Leslie L. Hughes, and Joseph E. Sheldrick for important contributions to the production of the book.

The author and editor appreciate very much the generous assistance of these people. In addition, we wish to thank each other for a rewarding collaboration, marked by stimulating discussion, excellent cooperation, and mutual understanding despite our never having met. We believe that the book's favorable reception owes much to that happy association and a shared philosophy regarding honesty and clarity in writing.

RAYMOND L. MURRAY, *Professor Emeritus*
North Carolina State University, Raleigh, North Carolina

KRISTIN L. MANKE, *Technical Editor*
Pacific Northwest National Laboratory, Richland, Washington

Contents

Illustrations

Tables

Questions and Concerns About Wastes

Energy and the Environment

Today the world faces two major problems related to energy and the environment: potential scarcity and pollution.

The environmental movement of the 1960s called attention to the growing problem of pollution to land, air, and water both in developed and developing nations. Concern increased around the world about the storage and release of chemicals. In the 1970s and 1980s, the problem of chemical waste became prominent. People became increasingly aware also of radioactive wastes, those that emit radiation as they break up or "decay."

The Arab oil boycott of 1973 focused attention on the potential scarcity. Costs of petroleum increased significantly, contributing to economic difficulties in both advanced and emerging countries. Nearly 20 years later, the world was again reminded of the potential scarcity of energy when Kuwait, an important oil-producing country, requested assistance from the United Nations. The response was the Persian Gulf War of 1991. Ironically, the war resulted in Iraq's committing ecoterrorism by burning and dumping 250 million gallons of oil, severely polluting land, air, and water.

The U.S. energy policy is to minimize oil vulnerability, to improve efficiency of energy use, to stimulate economic growth, to help developing nations, to encourage alternative sources to fossil fuels, and to protect the public and the environment. Nuclear power is viewed as one choice to meet national needs, along with natural gas, coal, and solar energy.

The Earth Summit Conference in Rio de Janeiro in June 1992 and subsequent conferences (see References) called attention to the challenging global problems of resource transfer, sustained development, and environmental protection. The Kyoto Protocol (see References) highlighted concerns about climate change resulting from the release of greenhouse gases from human activities, especially from coal-fired power plants.

Questions About Nuclear Energy

Although people recognize the need for various energy sources, they often raise the question about nuclear energy, "What is being done with radioactive wastes?" The federal government and the nuclear industry have stated that such wastes were known to be dangerous ever since they first were generated in large quantities in World War II, and that special care has been taken over the years to protect the public. Observers note, however, that decisions about final disposal of wastes have had a lower priority than production of nuclear weapons components in the overall nuclear development cycle. Examples of poor strategy and uncertainty in the waste management program are cited. The search for suitable disposal sites has stimulated public concern. News of disagreement among different branches and levels of government has led many to conclude that we do not know what to do with radioactive wastes.

The question "Is nuclear power safe?" is also raised frequently. People know of the effects of the atom bomb. They know that a nuclear weapon is not the same as a nuclear reactor, but they tend to associate the two and are uneasy because both involve fission and radioactivity. Many people think of all radiation as mysterious and harmful. Many are uncomfortable, worried about, or frightened by nuclear energy. Others, however, recognize that nuclear power is a clean source of needed electricity and support its continued or expanded use. Most people are aware that there are few fatalities from the use of nuclear energy. The Chernobyl reactor accident of 1986, however, heightened fear of nuclear power throughout the world. People are aware of greater dangers but feel that in those situations they are in personal control of their safety. The nuclear hazard is less acceptable than other, more familiar risks such as riding in automobiles.

Scientists are expressing alarm about damage to forests and inland waters and a possible global temperature rise because of emissions from burning fossil fuels. For this reason, the question also is heard "Does nuclear power help solve the problems of atmospheric pollution, acid rain, and the greenhouse effect?"

Finally, the terrorist attacks of September 11, 2001, raised concern about the vulnerability of all of the country's industrial and commercial facilities, including nuclear power plants.

The Need for Information

Approval for new nuclear electric plants and continued use of existing ones may depend on satisfactory demonstration of safe waste disposal. In making decisions about power plants, it is important that citizens and lawmakers alike know the nature of the waste problem and be able to distinguish opinions, feelings, and myths from the facts.

Unfortunately, too little public information on the subject of radioactive wastes is useful. Much of that said or written is rhetoric intended either to frighten or soothe. Debaters exaggerate to try to convince rather than inform. Polarization is such that neither side of the argument is credible to the average person.

The public is often confused by conflicting statements about nuclear energy and the waste problem. Scientists of presumed equal qualifications, for example, Nobel Prize winners, are seen taking opposite stands. There is a wealth of technical literature, but it is written for use by scientists and engineers familiar with the technical terms and background. Reports on plans and progress are in formal governmental language, which is often hard to translate into ordinary English.

This writer hopes to accomplish several things: (a) explain clearly the origin and nature of nuclear by-products, (b) provide facts and figures about nuclear wastes and the actions being planned on a national basis, (c) provide perspective on the safety of waste isolation systems, and (d) distinguish knowledge from opinion whenever possible, in an unbiased and candid manner. The author rejects exaggerated statements about the waste problem at both poles of the debate—assertions by proponents that it is merely a matter of politics, by opponents that a technical solution is impossible.

Understanding and Decisions

The most important premise behind this book is that an informed public will make the best decisions. We intend to help the reader understand such nuclear topics as uranium, radioactivity, radiation, and fission, along with the role of materials, chemical processes, and geology in the treatment and long-term handling of wastes. We will touch on people's knowledge and attitudes about radioactive wastes.

The nuclear debate may be characteristic of our times. Some say, "Government bureaucracy and industry try to go ahead with no interference by the public, but fortunately there are public interest groups to help the people speak out." Others counter, "We have dedicated leaders seeking to serve the public most effectively and economically, versus a small number of activists who would like to see power decentralized and the present system dismantled." Perhaps there is some truth in both views. The author believes that all thinking citizens have some interest in and concern about the waste problem and that a thorough explanation will be useful.

Some questions we hope to answer to the reader's satisfaction are:

1. What is radioactivity?
2. How are radioactivity and radiation related?
3. What are the effects of radiation on living beings?
4. How can we be protected from radiation?
5. In what ways is radiation useful or beneficial?
6. What are radioactive wastes?
7. Where do most of the radioactive wastes come from? How much do we have on hand and how much is produced each year?
8. What is the difference between high-level and low-level wastes?
9. How are nuclear wastes similar to or different from poisons from chemical processes? From emissions of electrical power plants using coal or oil?
10. What are nuclear power plants doing with their used fuel?
11. Why do we not yet have a place to dispose of spent fuel? When will a repository be available?
12. How are used nuclear reactor fuel and processed nuclear wastes transported? What precautions are taken to prevent accidents and to protect the public?
13. How vulnerable to terrorist attack are wastes being stored or transported?
14. What are the best ways to dispose of wastes?
15. What laws and regulations apply to the management of wastes?
16. What organizations are responsible for handling wastes?
17. How much space is needed to hold wastes? How far from public areas should disposal sites be located?
18. What kinds of rock are suitable for waste disposal? What do we need to know about rock formations?
19. How much does it cost to dispose of wastes? Who pays for it?
20. Is it safe to dispose of waste as planned?
21. What will be done with nuclear weapons materials as disarmament proceeds?
22. Is the problem of nuclear waste disposal overemphasized compared with other national problems?

The Sciences Help Explain

What one should learn so as to understand both the questions and the answers depends on a person's attitude. No information is needed if one says either "I am satisfied that those in government and industry will make good decisions" or "I am convinced that nuclear power is unsafe and must be abandoned at once." At another extreme, if one says, "I must have the same background as scientists and engineers who are advising on wastes," years of specialized college training would be required.

We take a middle ground and assume that the reader wishes to gain enough technical information to be able to think clearly about the subject and discuss it rationally. We believe it is helpful to know some atomic and nuclear science, to have been exposed to a little of the history of nuclear energy development, and to have some appreciation of what a nuclear reactor is and does. This background helps in learning about radioactive wastes and their management, and in understanding the issues that still remain to be resolved before the waste problem is put to rest.

Some basic science background is needed because many subjects bear on nuclear wastes: general science, chemistry, physics, biology, and earth science. Let us see how these sciences come into play:

- Wastes consist of many chemicals, composed of about half of the 113 discovered elements in the periodic table. Chemistry tells us how well we can remove certain hazardous substances, what form the wastes are in, what they will mix well with, how easily they are dissolved, and how fast they move through air, water, and earth.

- Physics describes properties of the nucleus of the atom, including how many types of radioactive atoms there are, how long they last, and what radiation they emit.

- Biology explains how radioactive materials may be brought to humans through food chains involving plants and animals, which organs of the human body have an affinity for certain materials, and how radiation damages cells and tissues.

- Earth science describes the features of rocks and soil in which wastes may be placed, explains how water is transferred through the ground, and identifies the types of media and best locations for long-term waste storage.

- Meteorology tells where airborne nuclear wastes might go and what might be the effects of emissions of other energy sources besides nuclear.

In the next several chapters, we review our knowledge of the atom, the periodic table of elements, simple chemicaMichetention is given to the effects of radiation and ways to protect human beings. We need such background for an understanding of radioactive waste.

CHAPTER 2

Atoms and Chemistry

Some Distinctions: Atomic and Nuclear

The scientific basis of radioactive waste management involves two levels: *atomic* and *nuclear*. These words are often confused; they refer, however, to quite different realms of matter. Atomic processes are chemical in nature, involving electrons of atoms or molecules. Nuclear processes are very energetic ones involving the inner core of the atoms.

Thus chemistry reveals what treatment or processing of used (spent) fuel from nuclear reactors will separate the useful from the useless. It tells what wastes might combine well with and tells what materials will make good containers. It also indicates how to prevent waste-particle migration in the ground.

On the other hand, the subject of nuclear physics explains the processes of radioactivity, the behavior and effects of radiation, and the production of new species of material by neutron bombardment in a reactor.

To understand the radioactive waste problem, we need both atomic and nuclear concepts. Let's review, very briefly, the important facts as discovered in the last hundred years or so.

Our World of Atomic Chemistry

Recall from basic science courses that all matter is made up of a small number of different kinds of atoms, the chemical elements. The first and lightest of these is hydrogen, and the ninety-second and heaviest natural element is uranium. There are 21 manmade elements that are heavier than uranium.

Chemical elements combine to form new substances called compounds. Some familiar processes, expressed as equations, are these:

burning of hydrogen:
    hydrogen + oxygen → water

preparation of table salt:
    sodium + chlorine → sodium chloride

rusting of iron:
    iron + oxygen → iron oxide

burning of natural gas:
    methane + oxygen → carbon dioxide + water

neutralization of an acid by an alkali (lye):
    nitric acid + sodium hydroxide → sodium nitrate+ water.

More complicated still are the reactions in living organisms, such as photosynthesis in plants and digestion in animals. The regularity with which chemical reactions take place led to the discovery of the atomic theory—that matter is composed of individual particles called atoms, which combine to form molecules. Atoms are extremely small. It would take a row of 10 million hydrogen atoms to span the head of a pin. One teaspoon of water contains an enormous number of molecules—around 200,000,000,000,000,000,000,000 (or 2×10^{23}).

Inside the Atom

Until the twentieth century, the internal composition of atoms was not known. Only after an experiment by Rutherford in 1911 was it realized that the electrically neutral atom had a central core (nucleus) of positive charge and an outer region of negative charge. Then studies by Bohr in 1913 revealed the relationship between light and atomic structure. He assumed a motion of the electron about the nucleus similar to that of a planet around the sun.

We will use that analogy to explain features of the atom. Picture the atom as a miniature solar system. In place of the sun, there is the heavy, positively charged central core of the atom, called the nucleus (plural: nuclei). This core is composed in general of still more basic particles—protons and neutrons. In place of the planets, there are electrons, the small, negatively charged particles that give us electricity.

The sketch representing an atom of hydrogen shows its single electron in orbit about the nucleus, which in this case is the proton. The proton has a weight about 1800 times that of the electron. In place of gravity, the force of electrical attraction holds the particles together.

As many readers know, the periodic table of chemistry lists the elements according to their chemical similarity. Each element is assigned a symbol and an "atomic number," which is the number of protons in the atom. On the atomic number scale (represented by the letter Z), hydrogen (H) is 1, helium (He) is 2, oxygen (O) is 8, iron (Fe) is 28, gold (Au) is 79, and uranium (U) is 92. Thus we say, "The atomic number for iron is 28," or more simply "For Fe, Z is 28."

How atoms produce and absorb light also comes from Bohr's theory. Electrons have the ability to suddenly "jump" to an orbit with a

The hydrogen atom. One electron is in orbit about the nucleus (proton). The atom is a hundred millionth of a centimeter across. If the proton were the size of a golf ball, the electron would be 2000 feet away.

smaller radius, accompanied by the emission of energy in the form of light. Similarly, an electron goes to a larger orbit when light is absorbed. If enough energy is supplied to the atom, the electron can be removed completely, leaving the positively charged nucleus, the proton (which is also the ion, H^+).

The next most complicated chemical element is helium, used as a gas in dirigibles like the Goodyear "blimp." It has two electrons in orbit and a nucleus with two protons and two neutrons. The neutron is an electrically neutral particle weighing almost the same as the proton. Diagrams here illustrate the arrangement of particles in the helium atom and in the much more complex uranium atom.

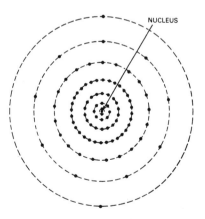

The helium atom. There are two electrons in orbit about the nucleus, which consists of two protons and two neutrons. The nucleus of helium is the same as an alpha particle.

The uranium atom with its 92 electrons in orbits around the nucleus. (Courtesy of Raymond L. Murray and Grover C. Cobb, *Physics: Concepts and Consequences,* Prentice-Hall, Englewood Cliffs, New Jersey, 1970.)

Radiation and Isotopes

It is important to note that the electron is the same as a *beta (β) particle*, one form of radiation (emitted in a nuclear reaction). The nucleus of a helium atom, He^{++}, is the same as an *alpha (α) particle*, another form of radiation. The third main type of radiation is *light*, which takes several forms. We see by means of ordinary visible light, which comes from atoms and molecules. The x-rays used for medical examination are more energetic than ordinary light. They can come from atoms or an x-ray machine. Gamma rays have still more energy and arise in the nucleus of the atom.

Each chemical element is made up of several types of atoms called isotopes. The difference between isotopes lies in the weight of their nuclei, which is determined by the number of protons plus neutrons.

For example, hydrogen is made up of the three species, as illustrated here: ordinary hydrogen with the proton as nucleus, deuterium with a proton plus neutron as nucleus, and tritium with a proton plus two neutrons as nucleus. We let A, called the "mass number," represent the number of nucleons (protons plus neutrons) in the nucleus. Thus A = 1, 2, and 3 for the isotopes of hydrogen. Recalling that Z = 1 for hydrogen, we can write symbols to uniquely specify them: ^1_1H is ordinary hydrogen, ^2_1H is deuterium or heavy hydrogen, and ^3_1H is tritium. The superscripts are mass numbers (nucleons); the subscripts are atomic numbers (protons).

The element hydrogen is familiar to us as one component of water, H_2O. Deuterium is rare in nature, there being only one atom for every 6700 atoms of ordinary hydrogen. Tritium is a manmade isotope. These heavier species of hydrogen are the ingredients in the fusion process being developed as a practical energy source.

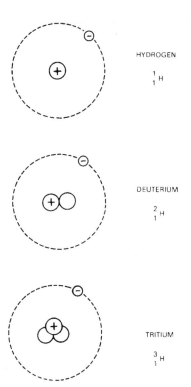

The three isotopes of hydrogen. Each has one proton in the nucleus; the difference between them is only in the number of neutrons in their nuclei—none, one, and two.

At the heavy end of the periodic table of natural elements is uranium, atomic number 92. It is composed of two main isotopes, mass numbers 235 and 238, with symbols $^{235}_{92}U$ and $^{238}_{92}U$. Only 0.7 percent is the lighter isotope; 99.3 percent, the heavier. Each of these two isotopes plays a role in the fission nuclear reactor. A great variety of chemicals appear in the operation of nuclear processes and devices, as we will take up later.

We have presented the abbreviated notation for isotopes because it is found in the technical literature. For most purposes, however, it is sufficient to use only the name and mass number, for example, "uranium-235" or "^{235}U."

CHAPTER 3

Radioactivity

The Process of Decay

Many of the isotopes found in nature are stable, meaning that they never change. Other isotopes, both natural and artificial, are radioactive, meaning that they are unstable and can change into another form.

Radioactivity is a process in which a nucleus spontaneously disintegrates or "decays." For the simplest example, see the diagram depicting the decay of hydrogen-3 or tritium. One of the two neutrons in its nucleus changes into a proton and an electron. The new nucleus, composed of two protons and a neutron, is the same as that of an isotope of helium (helium-3). The high-speed electron emitted is called a beta particle. Radioactivity can be described by a reaction equation. That for the decay of tritium is

TRITIUM   HELIUM-3   ELECTRON

$$^3_1H \longrightarrow \ ^3_2He \ + \ ^{\ 0}_{-1}e$$

The radioactive decay of the isotope tritium, shown as a diagram and by a word equation.

tritium → helium-3 + electron.

A few other important nuclear reactions that involve radioactive decay are:

uranium-238 → thorium-234 + alpha particle

cobalt-60 → nickel-60 + beta particle + gamma rays

iodine-131 → xenon-131 + beta particle.

Half-Life and Activity

One cannot predict when a particular nucleus will decay. If we could watch a group of radioactive atoms, some would decay at once, others later, still others much later. The number that decay in any second of time depends on only two things: how many there are and the nuclear species. Each isotope has its own "half-life" (t_H), which is the time it takes for half of any sample to decay. For instance, t_H for tritium is 12.3 years. Thus if we started now with 1000 atoms of tritium, after 12.3 years we would have 500 atoms, after 24.6 years, 250 atoms, and so on. The graph shows the trend with time of the number of atoms.*

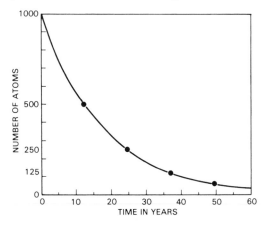

Graph of radioactive decay of tritium with time, starting with 1000 atoms. The half-life of tritium is 12.3 years.

Disintegrations of Nuclei

The rate of decay is called the *activity*, which is the number of disintegrations per second (d/s). The activity decreases with time in the same way that the number of atoms present decreases. Thus the hazard from the radioactive emissions decreases with time. In summary, the smaller the amount of material we have and the longer its half-life, the smaller is the activity and the safer is the sample. One rough rule when dealing with small quantities of material is that it takes 10 half-lives to eliminate a radioisotope. This rule is adequate when dealing with small quantities of material, because a fraction $(1/2)^{10} = 1/1024$ of the original atoms remains. However, when dealing with certain materials and situations, this rule is too crude to be of value, especially when factors of 1/1,000,000 or better are needed.

---

*It is easy to calculate the fraction f of particles that remain at any time, t. Let p stand for the number of half-lives elapsed, which is the ratio t/t_H. Then

$$f = \left(\frac{1}{2}\right)^p.$$

If the power p is an integer, we can do the calculation in our heads. For example, if the time is three half-lives, $t=3t_H$, then

$$p=3 \text{ and } f \text{ is } \frac{1}{8}.$$

If p is not an integer, we need to use a pocket calculator. For example, if $t_H = 12.3$ years as for tritium and t is 100 years, then

$$p = 100/12.3 = 8.13 \text{ and } f \text{ is } 0.00357.$$

Natural Radioactivity

Several of the heavier isotopes in nature are unstable and decay with the emission of an alpha particle. A good example is radium, which becomes the gaseous element radon when it decays, according to

radium-226 → radon-222 + alpha particle.

The half-life of radium is 1599 years and the activity is around 3.7×10^{10} d/s per gram. To measure the activity or "strength" of an unstable element, two different units can be used. The first is the curie*, defined as 3.7×10^{10} d/s. It was named after Marie Curie, who first studied the radioactivity of uranium and radium. When we say that a cobalt-60 radiation source has a strength of 1000 Ci, it means that the decay rate is $(1000)(3.7 \times 10^{10}) = 3.7 \times 10^{13}$ d/s. A more modern unit of activity is the becquerel (Bq), named after Antoine Becquerel who discovered radioactivity in uranium salts, which is 1 d/s. The strength of our cobalt source is thus 3.7×10^{13} Bq.

Natural Decay Chains

The radium-radon step is but one in a long chain of radioactive processes starting with uranium-238, as shown in the following table. This radioactivity is important because it occurs in the residues called "tailings" from the mining and milling of uranium ore. The final isotope is seen to be stable lead-206. Other natural chains start with uranium-235 and thorium-232.

Some isotopes decay with the release of an alpha particle only; others yield beta particles only; but many give both beta and gamma rays. The latter are similar to visible light except that their energy is much higher and they are better able to penetrate matter.

Uranium as a mineral found in nature is only mildly radioactive, because the half-lives of both its isotopes are very long: for uranium-238, 4.47 billion years; for uranium-235, 704 million years.[†] However, uranium is found in many types of rock, and because several of its descendants are more radioactive than uranium-238, many building materials give us a significant level of radiation. Radon-222, with half-life 3.82 days, is a noble (i.e., chemically inert) gas that enters buildings from the ground. It decays into radiation-emitting isotopes of polonium and bismuth, as seen in the table. Another naturally occurring radioactive element is potassium. Accompanying stable potassium-39 is weakly radioactive potassium-40 (0.0117 atom%) with a half-life of 1.26 billion years.

---

*Fractions of a curie are the millicurie (10^{-3}), microcurie (10^{-6}), nanocurie (10^{-9}), and picocurie (10^{-12}). For example, one nanocurie is 37 d/s.

[†]From a gram of pure uranium-238, the number of disintegrations per second is 12,400, corresponding to about 0.3 microcurie.

The Chain of Natural Radioactivity Starting with Uranium and Ending with Lead

(Reference: "Table of the Isotopes," Norman E. Holden,
in *CRC Handbook of Chemistry and Physics*, David R. Lide, Editor)

Isotope	Name	Half-Life*	Main Radiations Emitted
$^{238}_{92}$U	Uranium	4.44×10^9 y	α
$^{234}_{90}$Th	Thorium	24.1 d	β,γ
99.87% $^{234}_{91}$Pa^m 0.13%	Protactinium	1.17 m	β,γ
$^{234}_{91}$Pa	Protactinium	6.69 h	β,γ
$^{234}_{92}$U	Uranium	2.45×10^5 y	α
$^{230}_{90}$Th	Thorium	7.54×10^4 y	α
$^{226}_{88}$Ra	Radium	1.60×10^3 y	α
$^{222}_{86}$Rn	Radon	3.82 d	α
99.980% $^{218}_{84}$Po 0.020%	Polonium	3.04 m	α
$^{214}_{82}$Pb	Lead	26.9 m	β,γ
$^{218}_{85}$At	Astatine	1.6 s	α
99.979% $^{210}_{83}$Bi 0.021%	Bismuth	19.7 m	β,γ
$^{214}_{84}$Po	Polonium	164 μs	α
$^{210}_{81}$Tl	Thallium	1.30 m	β,γ
$^{210}_{82}$Pb	Lead	2.26 y	α,β
~100% $^{210}_{83}$Bi 0.00013%	Bismuth	5.01 d	α
$^{210}_{84}$Po	Polonium	138 d	α
$^{206}_{81}$Tl	Thallium	4.20 m	β
$^{206}_{82}$Pb	Lead	Stable	—

*Note that numbers are rounded off to three digits (y = year, d = day, m = minute, s = second).

   Among the hundreds of radioactive isotopes (also called radioisotopes or radionuclides) are found half-lives ranging from a small fraction of a second to billions of years, as we saw in the table. A stable substance has a half-life of infinity, of course. Each isotope has its own half-life, which is unaffected by any chemical treatment.

   We have emphasized the "natural" radioisotopes, i.e., those found in nature. "Artificial" radioisotopes can be produced by irradiating (bombarding) stable nuclei with various particles such as protons, neutrons, alpha particles, and deuterons. Radioisotopes also are by-products of nuclear fission, as will be seen later. The figure below, showing some nuclear reactions, provides information on neutron absorption and radioactivity, for use in later sections.

NEUTRON ABSORPTION

RADIOACTIVE DECAY PROCESS

Some nuclear reactions. The symbol D is often used for deuterium (heavy hydrogen $^2_1$H); the wavy line with arrow is intended to suggest that the gamma ray is an electromagnetic wave; half-lives of the decay processes are listed.

CHAPTER 4

Kinds of Radiation

Particle Energy

Three important types of radiation (α, β, and γ) were introduced in the preceding section. Recall that the beta particle is the same as the electron, which is a negatively charged particle with very small mass, found in atoms. The beta particle is emitted from the nucleus in radioactive decay also. An example is the reaction

strontium-90 → yttrium-90 → electron.

Strontium-90 is an important nuclear waste because of its relatively long half-life of 29.1 years.

The energy of a beta particle depends on the process from which the particle comes. Also, its energy determines its ability to penetrate matter and cause radiation damage.

The typical unit in which particle energies are expressed is the "electron volt." We can understand the unit by doing an experiment. Picture a 1-volt battery connected to two plates of opposite polarity. If an electron is carried across the gap, the energy given to the electron is said to be one *electron volt*, abbreviated *eV*. Now suppose that one electrically accelerates a beta particle to the maximum energy it has on emission from strontium. This time the energy is much larger— 546,000 eV or 0.546 MeV (million electron volts). These strontium-90 beta particles are still less energetic than some, e.g., those of about 4 MeV from gallium-66 and silver-112.

The electron volt as a unit of energy. One kilowatt-hour of energy, the amount used by a 1000-watt appliance in an hour, is equal to 2.25×10^{25} electron volts.

19

Remember that the alpha particle is the helium nucleus. It is emitted by reactions such as the decay of uranium-238 into thorium-234. The alpha particles have an energy of 4.2 MeV, typical for alphas emitted by heavy natural radioactive isotopes. A neutron is produced by certain reactions involving the bombardment of a nucleus of an element such as beryllium by alpha particles from radium decay. The final nuclear reaction is

beryllium-9 + helium-4 → carbon-14 + neutron.

Such processes gave rise to the neutrons that were used in the discovery of fission. However, the neutron is not involved in the radioactivity of nuclear wastes as practically no nuclei decay with neutron emission.

On the other hand, neutrons from outer space continuously produce radioactive carbon by the reaction

neutron + nitrogen-14 → carbon-14 + proton.

The radioactive carbon-14 is used to find the age of archeological items, as noted in Chapter 10. Neutrons can also produce new elements. Neutrons absorbed in uranium-238 give rise to the artificial (manmade) isotope plutonium-239. Plutonium-239 is an alpha particle emitter that can serve as nuclear reactor fuel or as a weapon, as we shall see later.

Gamma Rays

The gamma ray can be imagined to be a burst of energy, a particle, or a wave. Each view is correct in some sense. Gamma rays are at the high-energy end of what is called the electromagnetic spectrum. As shown in the chart here, it includes radio, microwaves, infrared, visible light, ultraviolet, x-rays, and gamma rays, in increasing order of energy.

For nuclear purposes, though, it is often better to think of a gamma ray as a particle of light or a bundle of energy called a photon. Many radioisotopes are gamma emitters. For example, cobalt-60 gives two gamma rays of energy around 1.25 MeV that can be used in medicine to sterilize equipment, conduct diagnosis, or treat cancer. Gamma rays are emitted at the same time as beta particles from many of the radioactive by-products of the fission process. Such substances are part of the radioactive waste as discussed in Chapter 7.

Radiation Spreading and Stopping

Let us now study the behavior of these various particles in air or in a vacuum. If we were in a laboratory with a small piece of radioactive material, say strontium or cobalt, the particles would come out in all directions. As seen in the sketch, radiation from such a source is similar to that from a light bulb. First, the rays become less intense

The electromagnetic spectrum showing that radio-TV, visible light, and gamma ray nuclear radiation are all waves.

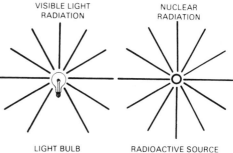

Comparison of the spreading of two forms of radiation—visible light and nuclear radiation. If the distance from the source is doubled, the intensity is divided by four.

with distance because of "inverse square spreading."* Prolonged exposure to radiation of either kind can be harmful. Infrared rays can burn the skin; the ultraviolet light component can damage eyes and cause skin cancer. Alpha particles, beta particles, or gamma rays can damage body tissue as well. Extensive data on the interaction of radiation with matter are provided by the National Institute of Science and Technology (see References).

---

*The intensity varies inversely as the square of the distance because all radiation goes through spheres of area $4\pi r^2$.

We can protect ourselves by backing away from the sources. Or, as shown in the sketch, we can interpose some solid material between us and the source of radiation. Each particle is slowed or stopped by collisions with atoms of the substance—in some cases by interactions with the electrons, and in others with the nuclei. A sheet of paper would be enough to stop alphas, but it would take a sheet of aluminum metal 1/25-inch thick to stop betas, such as those from strontium-90. These particles are said to have a certain "range," which increases with particle energy, of course.

Stopping radiation by various shields.

The intensity of gamma rays is merely reduced by passage through matter, just as light intensity is reduced by fog. A half thickness "t_H" can be defined, reminiscent of half-life. It is the distance it takes to cut the intensity of gamma radiation in half. For the cobalt gammas, the element lead has a t_H of close to 1 centimeter (cm), making it a useful shield for medical radiation diagnosis or treatment.

How radiation interacts with matter on a submicroscopic level is well understood. We can visualize, for example, the effects of a collision between a gamma ray and a simple atom, as sketched here. There are three possible events: (a) scattering, in which the gamma ray bounces off the electron; (b) ionization, in which the energy of the gamma ray goes into removing the electron from the atom, leaving an ion; and (c) pair production, in which the gamma ray energy is converted into the mass of two particles—an electron and a positron (positively charged electron). This process, pair production, illustrates Einstein's theory that energy and mass are two forms of the same thing.

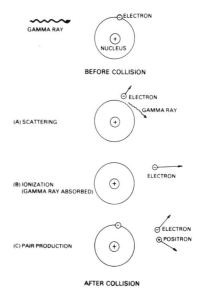

Effects of gamma radiation. There are three possibilities—scattering, ionization, and electron-positron pair production.

CHAPTER 5

Biological Effects of Radiation

Radiation Takes Many Forms

Modern people are already familiar with quite a few forms of radiation. They make use of visible light, both natural and artificial, in all their activities. They enjoy microwave ovens, radios, and televisions, which operate on low-frequency electromagnetic waves. They have experienced the well-known form of radiation exposure, sunburn.

Less familiar but real is the unfelt but continued bombardment by radiation from space and the earth. Also little understood are the radiations from nuclear devices and products. In this chapter, we will study the effects of rays and particles from radioactive materials, with special attention to low-level radiation.

Radiation and Living Cells

When high-speed particles such as alphas, betas, gammas, or neutrons strike living tissue, they slow down and stop just as if they had hit paper or aluminum or lead. The energy of motion of the individual particles is imparted to the biological cells as localized heat. Molecules in the cells are changed, or atoms are converted into ions, or the atomic nuclei are displaced from their positions.

The disruptive action caused by absorbed energy impairs human cellular functions. If the amount of radiation received is extremely small, there may be no significant damage. If the amount is very large, radiation sickness, genetic effects, or death may result. The words "large" and "small" are, of course, relative, so we need to express the amounts of radiation received in terms of numbers.

The biological effect of radiation, called dose or dosage, depends on the amount of energy absorbed and also on the type of radiation. Each type of radiation has a different effect on tissue. For example, in comparison with the effects of x-rays, gamma rays, and low-energy beta particles, neutrons are 10 times as damaging and alpha particles are 20 times as damaging. These factors that represent biological effectiveness are taken into account when the dosage is expressed in *rems*.* The rem is a unit of dosage just as the inch is a unit of length. For low radiation levels encountered regularly by human beings, the millirem (mrem) is used, that is, 1/1000 rem.

Mankind throughout all existence has been exposed to radiation from natural sources, including the ground, radionuclides in the body, and from space in the form of cosmic rays. Those rays bombard our atmosphere and all beings on the earth's surface. Another source that has been recognized as significant is the element radon. As noted in Chapter 3, radon-222 is the result of decay of radium-226 and is the parent of some radioactive bismuth and polonium isotopes. As a colorless, odorless, and invisible gas, radon seeps into buildings from the ground and is breathed by the occupants.

The diagram shows the average individual American's annual exposure, which is estimated to be 360 mrems. Doses and percents from the various sources are shown. The dose varies from one state to another because of differences in the uranium content of the earth. The dose varies with altitude, which affects the amount of screening effect of the atmosphere. The radioactive isotope potassium-40, as a component of the trace element potassium in foods, becomes a part of our bodies.

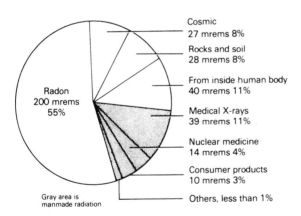

Sources of radiation exposure. The estimated annual dose to an average American is 360 millirems, according to National Council on Radiation Protection and Measurements Report No. 93, 1988.

*The actual energy absorption is measured in rads, with 1 rad as 0.01 watt-second energy absorbed per kilogram of tissue weight. The factors of biological effectiveness are applied to convert the number of rads into the number of rems. Modern scientific units are the gray (Gy), which is 100 rads, and the sievert (Sv), which is 100 rems.

Hazards to the Body

Questions often asked when there is extra radiation present are "How much danger is there?" and "What effect will the radiation have?"

There are two general classes of radiation effects—somatic, meaning damage to body tissue, and genetic, referring to hereditary characteristics. The main somatic effect is cancer such as leukemia or tumors. Genetic effects involve transmitted abnormalities or early death.

Each tissue and organ of the body has a different degree of sensitivity to radiation effects. The blood-forming tissue, the gastrointestinal tract, and the gonads are some of the more readily affected. The body's external layer of skin provides some protection from bombardment by alpha particles because the particles penetrate only a short distance. However, radioactive material that emits alpha particles can be very hazardous if taken into the body.

Evidence of these physiological effects of massive radiation dosage comes from many sources: experiments on laboratory animals such as mice; observation of side effects of radiation treatment of certain diseases; data from the past on radium poisoning of people who painted luminous watch dials; the incidence of lung cancers among uranium miners who worked with inadequate ventilation of radioactive radon gas; studies on the survivors of the atomic bombing of Hiroshima and Nagasaki; and, finally, survivors of the accident at the Chernobyl Nuclear Power Plant.

Studies of radiation effects lead to two conclusions. The first is that there is no direct evidence of genetic damage in humans, even though mutations have been observed in plants and lower animals. The second is that a single radiation dose of around 400 rems will be fatal to half of those who receive it, while half will survive, perhaps with some impairment of function. Because such large doses are rare, we are more interested in the effects of small doses.

Low-Level Radiation and the Linear Model

The subject of low-level radiation is subtle and complex. Let us start with some facts. First, there are no directly observable effects on human beings of a radiation dose smaller than 10 rems. Second, people are exposed continuously to a radiation background from cosmic rays and the ground of more than 200 mrems per year in the United States and several times larger in some countries such as India or Brazil, where there are large natural deposits of radioactive minerals. There is no evidence that the higher doses in certain geographic regions cause excessive health risk.

It is logical to assume that there is some level of dosage below which there is no permanent effect because of the body's ability to recover. On the other hand, it can be argued that any amount of radiation is harmful.

We are faced with a large gap in information between the extremes of zero radiation (and zero effect) and very high radiation doses. The simplest assumption is that radiation effect is directly proportional to radiation dose. Thus, as shown in the drawing, one starts with the available data on effects of high radiation doses and draws a straight-line graph down to zero. This solid line expresses an important statement, often called the "linear model" or "linear hypothesis." Advisory committees continue to recommend this simple model and the Nuclear Regulatory Commission implements it in its rules. The straight-line graph predicts a larger effect than if there were a threshold and thus is "conservative," i.e., is favorable to safety.

Such a stand has been criticized, however. The American Nuclear Society (see References) states, "there is insufficient scientific evidence to support the use of the Linear No Threshold Hypothesis (LNTH) in the projection of the health effect of low-level radiation." Cohen (see References) provides a number of reasons for the failure of the theory, including the fact that the human body has powerful repair mechanisms. Mitchel (see References) describes experiments with human and rodent cells and with animals that contradict the LNTH.

Many believe that there is negligible residual damage at low levels, implying a threshold in the dose-effect curve. Indeed, there is considerable support for the applicability of a process called "hormesis," where a small amount of a substance is helpful but a large amount is dangerous. A familiar example is aspirin. The lower U-shaped dotted curve may be more nearly correct.

Research continues on the effect of radiation on cultures of cells and on living organisms in the interests of both radiation protection and improvement in medical treatment.

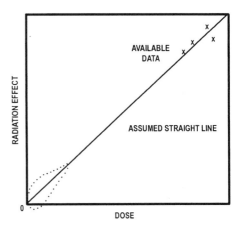

Possible effects of radiation dose. The upper dotted curve is considered unlikely. The solid straight line is a regulatory prescription. The lower dotted curve showing benefit at very low doses may be correct.

How Sure Can We Be?

Because we cannot observe the effect of low-level radiation on an individual, we must view the effect statistically. Thus if a large number of people all receive the same small dosage, it is presumed that there will be a few fatalities over the course of years. A unit of total population dose is called the person-rem. For instance, if each person in a group of 100 people received 5 mrems (0.005 rem), the population dose would be 100 x 0.005 = 0.5 person-rem. The linear model yields a prediction of 1 additional cancer death for each 2500 person-rems.

We can apply this to the Three Mile Island reactor accident of March 1979 near Harrisburg, Pennsylvania. It was estimated that the 2 million people living within a 50-mile radius received a total of 3300 person-rems. The predicted additional cancer deaths (over and above those from other causes) would thus be 3300/2500 = 1.32. Assuming that there would be an equal number of radiation-caused genetic deaths, the total comes out fewer than 3 additional fatalities attributable to the accident. Looked at in another way, each of the 2 million people in the Harrisburg area could say that his or her increased chance of dying as a consequence of the incident was about 1 in 750,000.

The Chernobyl reactor accident of April 1986 resulted in the exposure of a few people to high radiation doses, well above the lethal level. It also gave millions of people in Europe small increases over natural background. Based on LNTH, one would predict that thousands would eventually die prematurely.

The effects of various radiation doses over a wide range of values are shown in the diagram. The scale goes down by ten at each mark. For example, the dose at a nuclear plant boundary is lower by a factor of 100,000 from a fatal figure.

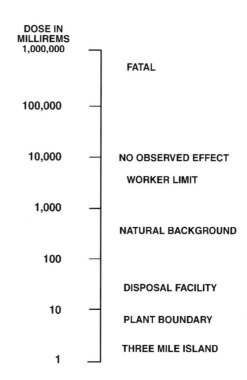

Examples of radiation doses. The regulatory limit for a disposal facility for used nuclear fuel is 15 millirems. Note that the scale is logarithmic.

What Is Risk?

We can understand risk better by examining data on fatalities from various causes, including diseases and accidents. The following table gives U.S. figures for the year 2000. Let us use that table to estimate the chance of death each year from some common causes. The number of fatalities from heart disease was 709,894 out of a total population of around 275,255,000. The chance of dying of heart disease each year is thus 1 in 388. From the figure on motor vehicle accidents, 41,804, the chance is 1 in 6584. Note that these calculations are averages over the whole population.

Clearly, one's chances of being killed on the highway greatly exceed those of being killed by radiation; but most people feel that they have control of their fate when they drive but do not if a nuclear reactor is operated by someone else. Also, automobile accidents are common, easily understood, and only a few people are affected in each accident, while reactor accidents are rare and mysterious, and it is possible that a large number of people could be affected. We know that many people deliberately take chances with their lives by smoking, drinking, and taking drugs. Others find excitement in hazardous sports. Also the perception of risk varies widely among different groups of people, as noted by Wilson and Crouch.* Whereas college students rank nuclear power first, experts put it at 20 out of 30.

The question is often asked, "Why do we not know the effect of low-level radiation more precisely?" The basic reason is that

Major Causes of Death in the United States for 2000

(From *National Vital Statistics Report*, Vol. 49, No. 12, October 9, 2001)

The total population was approximately 275,255,000

Heart disease		709,894
Cancer		551,833
Stroke		166,028
Lung disease		123,550
Accidents, total		93,592
Motor vehicle	41,804	
Other	51,788	
Diabetes		68,662
Pneumonia, influenza		67,024
Alzheimer's disease		49,044
Kidney disease		37,672
Blood poisoning		31,613
Suicide		28,332
Liver disease		26,219
Hypertension		17,964
Pneumonitis		16,659
Homicide		16,137
All other causes		400,401
Total		2,404,624

---

*Richard Wilson and Edmund A. C. Crouch, *Risk Benefit Analysis*, Harvard University Press, 2001, Table 4-1, p. 111.

there are too many competing causes of injury. Radiation can cause cancer but so can exposure to many kinds of foods, drugs, chemicals, and other pollutants. To demonstrate this statistical situation, suppose that we made a very careful study of 1000 atomic workers over their lifetimes and found that 234 of them died of cancer. The table shows that the number of such fatalities in an average group of 1000 deaths in the population would be 229. One might conclude from the above two numbers that 5 excess deaths are due to radiation exposure. However, statistical variations are expected among samplings of people. For example, there is a difference of around 4 cancer deaths per 1000 over the 6-year period 1995-2000, using the total of more than 2 million fatalities. In any actual sample of 1000 deaths, the fluctuations will be even greater. It is not possible to conclude that radiation caused any of, all of, or even more than the five extra fatalities. Another factor that clouds the issue is the "healthy worker effect," in which people in an industrial group studied tend to take better care of their health than average citizens. Although continued studies should and will be made, there is not likely to be a major improvement in accuracy of the estimated effect of low-level radiation.

The subject of risk has many dimensions, as aptly discussed by Wilson and Crouch. The number of deaths from various causes varies greatly with the ages of the victims. For example, young men experience more automobile fatalities while more elderly people die of cancer. More meaningful than mortality statistics are data on life-shortening. Cohen (see References) contrasts the loss from heavy drinking (11 years) with the most pessimistic view of the effect of nuclear power (2 days). Another measure of risk is the amount of money needed in a year to save one life. Tengs et al. (see References) tabulated a large number of "life-saving interventions," presumably taking account of all activities required to achieve the goal. Selected costs are:

Enforcement of a seat belt law $69

Chlorination of drinking water $3,100

Radon remediation in homes $140,000

It is clear where limited funds should be spent.

Comparison with Chemical Wastes

Radioactive wastes from various nuclear processes, one source of harmful radiation, should be viewed in perspective. Certainly the hazard seems very ominous because some of the isotopes last for thousands or even millions of years. Some chemical poisons, however, such as arsenic, lead, and mercury last forever. In a sense, radioactive materials are degradable, while nonradioactive poisons are not.

CHAPTER 6

Radiation Standards and Protection

When high-energy atomic and nuclear radiation were first discovered early in the last century, many experimenters received excessive doses of radiation. The discovery of x-rays led to widespread use of this radiation for medical diagnosis. Scientists and doctors were not aware of the biological damage that such radiation could produce and of the need for great caution in administering x-rays and handling radioactive materials. As a result there were a number of radiation burns and fatalities. Radioactive materials also became more common in industry. In painting the naturally radioactive element radium on watch dials to make them luminous in the dark, workers who pointed their brushes with their lips received excessive radiation and many died of leukemia. Today, radiation standards exist for workers and the public.

Standards for Protection

Over the years, however, more information became available and greater safety measures were developed. Certain official organizations were established and standards for maximum allowed exposure were set. The limits on dosage have generally come down. New sciences appeared: radiology, health physics, and radiation protection. In the early days, the test of radiation dose was reddening of the skin. Such primitive methods have been replaced by detection using sensitive electronic instruments and photographic film. There are well-known rules for ensuring the safety of patients who receive x-rays. We now have detailed tables of standard values of the limits on concentrations of radionuclides in air or water and information on the allowed dose to specific tissues and organs of the body. Organizations responsible for recommending new standards are the International Commission on Radiological Protection (ICRP), formed in 1928, followed the next year by the United States counterpart, the National Council of Radiation Protection and Measurements (NCRP).* The NCRP has specified dose limits for nuclear plant workers.

---

*The NCRP consists of 85 council members from universities, hospitals, national laboratories, and government agencies. The body is not an official government agency, however.

What Determines Radiation Dosage?

The radiation dose that we could receive depends in part on the amount of radioactivity in the air that we breathe or the water that we drink. Limits on the concentration of a radionuclide are obtained by use of the science of health physics. Several questions are considered:

- First, what is the affinity of the isotope for an organ of the body? It may be an element that deposits in the bone, such as strontium-90; it may deposit mainly in the thyroid gland, such as iodine-131 or iodine-129; it may be a gas that affects the lungs, such as radon-222 or krypton-85.

- Second, what radiations does it emit? If it gives off only soft (low-energy) betas, as carbon-14 does, the hazard is less. But if it emits alphas, as plutonium does, or hard (high-energy) gammas, as cobalt does, the hazard is greater.

- Third, what is its half-life? If it is extremely short, the isotope may be practically gone before the body takes it in. If it is very long, the activity is low and thus little radiation is received.

- Finally, how does the body react when the isotope is ingested? Certain heavy-element oxides, if taken orally, would not remain in the body but would soon be eliminated. If breathed as large particles, however, they might lodge in the lungs and stay for a long time. An isotope of a chemical ordinarily found in the body, such as hydrogen, carbon, oxygen, nitrogen, sodium, or chlorine, would become a part of the body but would also be eliminated rather soon because of normal body processes. Thus hydrogen-3 (tritium) and carbon-14 are readily eliminated as water and carbon dioxide.

The radiation effect of a radioisotope that has entered the body thus depends on how rapidly the substance is removed by a combination of radioactive decay and biological elimination. Each process has a half-life. A formula involving the half-lives of the two processes is used to find the "effective" half-life of the substance.* For example, the ordinary half-life for tritium is 12.3 years and its biological half-life is 12 days, but its effective half-life is only 11.97 days.

---

*If t_H is the radioactive half-life and t_B is the biological half-life, the effective half-life t_E is found from the formula

$$1/t_E = 1/t_H + 1/t_B.$$

Values of t_B are cesium-137, 70 days; strontium-90, 50 years; and plutonium-239, 200 years, as listed in ICRP Publication 2.

Radiation Protection Practices

The first of a sequence of documents known as the BEIR reports (see References) provides some general guidance on radiation protection standards. The principles in simplified form are as follows:

1. Allow no exposure unless there is an important benefit.

2. Protect the public but do not waste large amounts of money on small improvements.

3. Radiation risks should be small compared with normally accepted risks.

4. The average dose to many people should be much less than that for an individual.

5. Medical radiation exposure can and should be reduced by avoiding mass x-rays, by inspection and licensing, and by proper training and certification.

6. Cost-benefit analysis should be applied to the nuclear industry.

7. Extraordinary efforts should be made to minimize the risk of a serious reactor accident.

8. Occupational and emergency exposure limits should be set.

9. The relationship between radiation and ecology should be studied.

10. Good estimates and predictions should be sought.

Regulation of Radiation

For many years before 1994, regulatory limits on radioactivity in air and water were expressed as maximum permissible concentrations (MPCs). As of 1994, however, the Nuclear Regulatory Commission (NRC) required the use of two new measures. The first of these is the "annual limit of intake," abbreviated ALI. For example, for a nuclear plant worker, the ALI for oral ingestion of cesium-137 is 100 μCi and that for inhalation is 200 μCi. The second measure is the "derived air concentration" (DAC), being the activity in air that would lead to an ALI if breathed continuously during work hours. For cesium-137, this figure is 6 x 10^{-8} μCi/ml.

Values of the ALI and DAC as they vary with each radionuclide are tabulated in the publication *Code of Federal Regulations 10 Energy*,* Part 20, Standards for Protection Against Radiation,

---

*This book, reissued yearly by the Superintendent of Documents, includes regulations of the Nuclear Regulatory Commission, based on specifications by the Environmental Protection Agency. Sections of 10 CFR 20 can be accessed at the NRC Web site (www.nrc.gov).

abbreviated as 10 CFR 20. Limits are also given on the concentration of radionuclides above natural background in effluents of air and water from a nuclear facility. The table shows the maximum concentrations for some commonly found radionuclides. These figures are lower than those allowed for plant workers. For example, the air limit for cesium-137 is 2×10^{-10} μCi/ml, which is 1/300 of the occupational number. Note that the concentration limits for weak beta particle emitters tritium (^{3}H) and carbon-14 are much higher than those for energetic alpha-emitting neptunium and plutonium. It turns out that the chemical toxicity of uranium can exceed the radioactive hazard. If several isotopes are in the air or water, their radiation effects must be added.

Maximum Effluent Concentrations, μCi/ml, Above Background for Selected Radionuclides

(From: *Code of Federal Regulations 10 Energy*, January 1, 2002, Superintendent of Documents)

Isotope	Air	Water
Carbon-14	3×10^{-9}	3×10^{-5}
Cesium-137	2×10^{-10}	1×10^{-6}
Hydrogen-3	1×10^{-7}	1×10^{-3}
Iodine-129	4×10^{-11}	2×10^{-7}
Iodine-131	2×10^{-10}	1×10^{-6}
Krypton-85	7×10^{-7}	—
Neptunium-237	1×10^{-14}	2×10^{-8}
Plutonium-238	2×10^{-14}	2×10^{-8}
Plutonium-239	2×10^{-14}	2×10^{-8}
Radium-226	9×10^{-13}	6×10^{-8}
Radon-222	1×10^{-10}	—
Strontium-90	3×10^{-11}	5×10^{-7}
Uranium-235	3×10^{-12}	3×10^{-7}
Uranium-238	3×10^{-12}	3×10^{-7}

The NRC specifies in 10 CFR 20 the maximum annual occupational dose, taking into account the sensitivity to radiation of body organs and tissues. As described in the regulation,* organ-weighting factors are used to arrive at a "total committed effective dose equivalent," which considers both external and internal exposure. The basic limit is 5 rems per year, with somewhat higher figures allowed for the lens of the eye (15 rem/yr) and the skin, hands, and feet (50 rem/yr).

For the general public, the limits are considerably lower, being only 0.1 rem (100 millirem) per year. For members of the public living near a radioactive waste disposal site, the dose is limited still further, to 25 millirems per year. The reason for the difference between limits for workers and the public is that workers voluntarily accept employment where radiation may be found. On the other hand, workers are protected through monitoring and control practices

---

*10 CFR 20, Section 20.1003, Definitions.

such that their exposure seldom reaches the regulatory limits. There are special rules that further limit the dose for minors and pregnant women.

"As Low As Is Reasonably Achievable"

Any radiation program licensed by NRC, as spelled out in 10 CFR 20.1101 and 10 CFR 20.1003, "shall use to the extent practicable, procedures and engineering controls based on sound radiation protection principles to achieve occupational doses and doses to members of the public that are as low as is reasonably achievable (ALARA)." The acronym "means making every reasonable effort to maintain exposures to radiation as far below the dose limits . . . as is practical . . . taking into account the state of technology, the economics of improvements in relation to the state of technology . . . (and) to benefits to the public health and safety, and other societal and socioeconomic considerations, and in relation to utilization of nuclear energy and licensed materials in the public interest." The application of this principle led the NRC (in 10 CFR 50 Appendix I) to limit annual whole-body dose to any individual outside a nuclear power plant to 3 millirems via liquid and 5 millirems via air. The annual total release of radioactive material (excluding tritium and dissolved gases) should be no more than 5 curies.

The use of ALARA is seen to involve judgment based on information on both radiation and cost. The logic of the principle has three parts. First the regulatory dose limits must be met. Then efforts should be made to do even better. The NRC suggests the use of $1000 per person-rem in cost-benefit calculations (10 CFR 50 Appendix I). Finally, zero risk is not sought, as it would obviously entail infinite cost.

Nuclear plants meet the standards set by the NRC by using a combination of sensitive radiation-detection equipment and certain procedures. The photograph shows a plant worker using a portable radiation instrument. Radiation levels are measured by radiation detectors and dosimeters, while radioactive contamination is determined by samples of water or air that might contain radioactive isotopes. For improved accuracy of measurement, large water samples can be evaporated to increase the radioactive concentration, and large air samples can be passed through filters that collect the contained radioisotopes.

The values of the maximum concentrations in the foregoing table are extremely small, but the very fact that the isotopes are radioactive and emit radiation makes it possible for them to be detected. Even though radiation cannot be seen, felt, heard, smelled, or tasted, radiation detectors can sense its presence and the amount of hazard can be measured.

Radiation worker with protective clothing and dosimeter. (Courtesy of Nuclear Energy Institute.)

How To Protect Against Harm

Radiation hazard to the human body is of two types—internal and external, each with somewhat different methods of protection.

To protect against internal radiation, it is necessary to avoid ingestion of radioactive particles by eating, drinking, or breathing. Thus the source of radioactivity must be isolated or rendered harmless. Storing the material in a safe place until it decays solves the problem if the half-life is short, for example, iodine-131 with its half-life of 8 days. Releasing the material into very large volumes of air or water can dilute the activity to below the maximum concentrations shown in the preceding table. Mixture with dirt in the ground is possible in some applications.

To protect against external radiation, three factors can be used— distance, time, and shielding. A person is safer the farther from the source of radiation, the shorter the time of exposure, and the thicker the shielding. The following sketch shows some ways dosage is minimized. The radiation warning symbol is used universally; rope barriers remind workers of potential hazard. Exposure to radiation from wastes is prevented by use of protective containers and by isolation of the radioactive material, as discussed later.

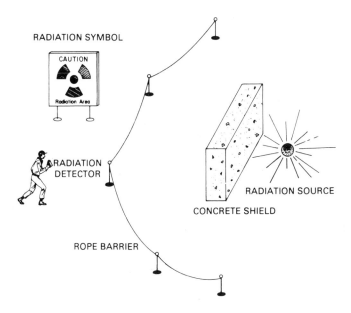

RADIATION SYMBOL

CAUTION

Radiation Area

RADIATION DETECTOR

CONCRETE SHIELD

RADIATION SOURCE

ROPE BARRIER

Examples of methods of protection against external radiation.

CHAPTER 7

Fission and Fission Products

The Splitting of the Nucleus

The radioactive wastes that are by-products of nuclear power generation arise mainly from the fission process. Fission is the splitting of a nucleus into two parts, triggered by absorption of a neutron.*

Using the sketch, picture the stages, beginning with a neutron approaching the nucleus of uranium-235. The neutron is absorbed and forms uranium-236, a compound nucleus. This particle is unstable, with some forces trying to hold it together, others trying to disrupt it. Imagine internal vibrations being set up that cause the shape of the nucleus to be distorted. Electric repulsion forces then cause the nucleus to separate into two fragments that fly apart at high speed. These fission fragments are the radioactive "fission products." Their energy of motion will eventually be converted into useful heat. As shown, several neutrons, gamma rays, and beta particles are released during this violent separation.

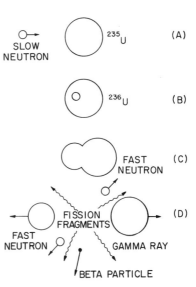

Stages in the fission process. A slow neutron approaches the uranium-235 nucleus in step (A), is absorbed to form uranium-236, as in (B). The nucleus becomes distorted as in (C), and the two fragments fly apart at high speed and various radiations are emitted, including neutrons, as in (D). (Adapted from Raymond L. Murray, *Nuclear Energy*, Butterworth-Heinemann, 2001.)

---

*In only a few isotopes such as californium-252 fission happens spontaneously (i.e., the isotope does not need to absorb a neutron).

Nuclei of uranium-236 may split in many different ways. If three neutrons were released, the most likely atomic weights would be around 140 and 93. The graph shows which fission product masses are more likely. It tells us, for example, that isotopes with A = 134 are produced in about 7 percent of the fissions. We see that masses 90 and 137 are quite abundant.

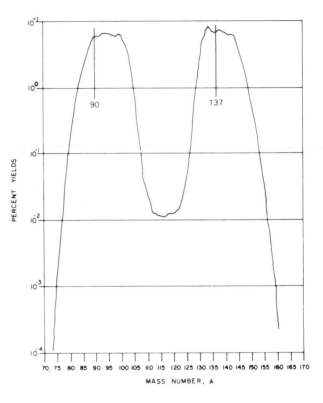

Yield of fission products according to mass number. Many elements and isotopes are present, both stable and radioactive.

Typical nuclear reactions are:

uranium-235 + neutron → uranium-236
uranium-236 → krypton-90 + barium-144 + 2 neutrons.

The uranium-236 splits into fission fragments in about 85 percent of the cases but merely releases a gamma ray in 15 percent. Note, although we talk about "^{235}U fission," it is really the ^{236}U that divides.

The terms "fissionable" and "fissile" are often used to describe types of nuclei. All nuclei can be made to undergo fission if the bombarding neutron has high enough energy. In other words, all nuclei are fissionable, but only a few, the fissile isotopes, can be fissioned with slow (thermal) neutrons. These are uranium-235, uranium-233, plutonium-239, and plutonium-241. Another technical term "fertile" refers to an isotope that can be converted into a fissile one by absorbing a neutron. Examples are uranium-238, thorium-232, and plutonium-240.

Several neutrons come from the fission process. Thus a chain reaction is possible, in which neutrons cause fissions that release neutrons that produce more fissions and so on, a process called "neutron multiplication." A quantity called the "multiplication factor" is symbolized by k. It represents how many neutrons are produced by one neutron on the average, and indicates whether an accumulation of material is subcritical, k less than 1; critical, k equal to 1 (self-sustaining); or supercritical, k greater than 1. Recall that "critical mass" is now a part of our vocabulary as anything of the right amount to be self-sustaining. When the size, shape, and materials are proper, we can have a steadily operating source of energy, a nuclear reactor.

The minimum ingredients are uranium and neutrons; many other components are needed to take away heat, provide structural strength, and permit control and safety. Uranium, through its fissile isotope uranium-235, is the fuel for a nuclear reactor; it is "burned" in the sense that the absorption of a neutron causes energy to be released.

Radioactive Fission Products

The fission fragments are individual nuclei. Collectively they are fission products, about 800 different isotopic species. Most of them are radioactive, emitting a series of beta particles and gamma rays, with half-lives that generally start out short and become longer in later stages of the chain.

Two important chains are depicted on page 42, with the half-lives for each decay shown. Note, as beta particles are released, the atomic number of the isotope increases. These chains lead to the nuclear waste isotopes strontium-90, half-life 29.1 years, and cesium-137, half-life 30.2 years. Some other important fission products and their half-lives are: iodine-131, 8.04 days; cerium-144, 284.6 days; cesium-134, 2.065 years; krypton-85, 10.73 years; technetium-99, 213,000 years; and iodine-129, 17 million years. Other isotopes produced are hydrogen-3 (tritium), 12.33 years, and carbon-14, 5715 years. Several fission product isotopes are valuable for use in research, medicine, and industry. New commercial applications such as food preservation by radiation treatment may increase the demand for such radionuclides.

When fuel is first installed in a nuclear reactor, it has a rather low radioactivity, because of long-lived isotopes of uranium of mass numbers 234, 235, and 238. Upon irradiation by neutrons, fission products are produced and decay, but they build up to a rather

$^{90}_{35}$ Br	bromine		$^{137}_{52}$ Te	tellurium
β ↓ 1.91s			β ↓ 2.5s	
$^{90}_{36}$ Kr	krypton		$^{137}_{53}$ I	iodine
β ↓ 32.3s			β ↓ 24.5s	
$^{90}_{37}$ Rb	rubidium		$^{137}_{54}$ Xe	xenon
β ↓ 2.6m			β ↓ 3.82m	
$^{90}_{38}$ Sr	strontium		$^{137}_{55}$ Cs	cesium
β ↓ 29.1y			β ↓ 30.2y	
$^{90}_{39}$ Y	yttrium		$^{137}_{56}$ Ba	barium
β ↓ 2.67d			(stable)	
$^{90}_{40}$ Zr	zirconium			
(stable)				

Two fission product decay chains leading to the important waste isotopes strontium-90 and cesium-137. These isotopes dominate the radioactivity in the waste for several hundred years. See also the following graph. (Data from *CRC Handbook of Chemistry and Physics*.)

constant inventory. The decay with emission of betas and gammas contributes to the useful heat energy of the reactor but also lasts for years after fuel is removed from the reactor. The graph shows the slow decline in rate of heat generation.

Heat generated by fission product isotopes as it depends on time after removal from a nuclear reactor. (After Bernard L. Cohen, *Scientific American*, June 1977.)

Energy from Fission

Each fission reaction releases a very large amount of energy, about 200 MeV, i.e., 200,000,000 eV. The fission fragments carry about 166 MeV of this total, and the neutrons, 5 MeV, and about 20 MeV goes to other radiations. The 200 MeV figure is to be contrasted with the low energy obtained by burning a chemical fuel. When hydrogen atoms react with an oxygen atom to produce a molecule of water, the energy yield is only 3 eV.

Per pound of fuel burned, the energy from nuclear fission is millions of times that from burning a chemical fuel such as coal or oil. Hence, the weight of nuclear wastes per unit of energy produced is extremely small. One gram of waste results from one megawatt-day of reactor heat energy production.* In contrast, there is a weight of some 2.5 tons of waste solids and gases from burning of a fossil fuel such as coal to produce the same amount of heat.

Fission is sometimes confused with another important nuclear process, fusion. The latter brings together two light isotopes such as deuterium and tritium to "fuse" or combine and release large amounts of energy. The fusion process, which is the source of the sun's energy, is being investigated as a potential source of practical energy. If successful technically, it would produce relatively few radioactive wastes. It will probably not contribute to the world's energy supply until the middle or end of this century.

---

*Recall that 1000 watts is a kilowatt and a million watts is a megawatt.

CHAPTER 8

The Manhattan Project

Wartime Wastes

Large volumes of nuclear wastes were produced in World War II in the effort to collect the plutonium needed for atomic bombs. This work was part of a defense effort with the code name "Manhattan Project." A review of its history* will reveal how the present waste situation arose.

We recall that direct involvement of the United States in the war began in December 1941 following the attack on Pearl Harbor by the Japanese. The Allies—consisting mainly of the United States, Great Britain, Russia, Canada, and France—were opposed by the Axis powers—Germany, Japan, and Italy.

The Atom Bomb

The Allies were particularly worried that Germany might develop and use a fission weapon. The Germans had discovered the fission process and had a high level of scientific and technical skill.

Urged by Albert Einstein and others, President Franklin D. Roosevelt launched a research and development program known as

The Manhattan Project: Research, Development, and Production

Location	Activities
Columbia University, New York City, New York	Gaseous diffusion uranium isotope separation
University of Chicago "Metallurgical Laboratory," Chicago, Illinois	First nuclear reactor; materials and chemical studies
University of California, Berkeley, California	Electromagnetic uranium isotope separation; chemistry of plutonium
Clinton Engineer Works, Oak Ridge, Tennessee	Production facilities; chemical process and reactor research
Hanford Engineer Works, Richland, Washington	Plutonium production reactor
Los Alamos, New Mexico	Weapons research
Iowa State College, Ames, Iowa	Reactor materials

---

*Comprehensive treatments of the project are in books by H. D. Smyth and by Richard Rhodes, and in a report by F. G. Gosling (see References).

the Manhattan Project, briefly outlined here. The ultimate objective of the whole enterprise was to harness the fission process to form an explosive, i.e., a bomb. The first material sought was highly enriched uranium, around 90 percent ^{235}U, starting with natural uranium, 0.7 percent ^{235}U (see the illustration). The second material was plutonium, the result of neutron bombardment of ^{238}U.

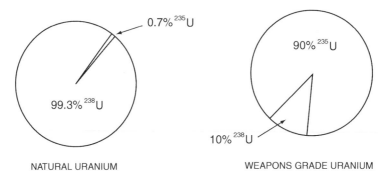

Composition of different enrichments of uranium.

Two types of weapons in which the materials were to be used were devised at Los Alamos in New Mexico, under the direction of J. Robert Oppenheimer. In the "gun" type, sketched here, a cylinder of enriched uranium was propelled at high speed down a tube by a chemical explosive. It entered a cavity hollowed out of another enriched uranium cylinder. The result was a supercritical condition. In the "implosion" type, also illustrated, chemical explosives compressed a small sphere of plutonium, bringing it to the supercritical state. In both types, there was an almost instantaneous release of enormous energy in the form of nuclear radiation and heat. A uranium gun

Nuclear weapons of the fission type. The top diagram is the gun-type uranium bomb. The bottom diagram is the implosion-type plutonium bomb.

bomb was dropped on Hiroshima. Plutonium implosion bombs were tested at Alamogordo, New Mexico, and dropped on Nagasaki, ending the war with Japan. In nuclear weapons, the chain reaction is used to cause rapid neutron multiplication through the consumption of fissile nuclei, as shown in the sketch.

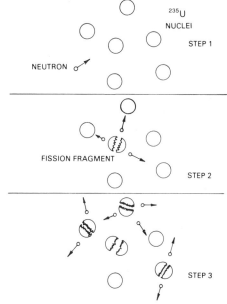

The chain reaction. In just three steps one neutron is "multiplied" to form seven neutrons. Four uranium nuclei are fissioned in this example, each giving rise to energy.

The separation of ^{235}U from ^{238}U was needed to provide a readily fissionable material. Techniques tested were thermal diffusion, centrifuge, electromagnetic, and gaseous diffusion. Research on the electromagnetic isotope separation process was conducted at Berkeley under Ernest O. Lawrence. Large-scale facilities at Oak Ridge produced enough highly enriched uranium (about 90 percent ^{235}U) for one of the first bombs. Gaseous diffusion, as illustrated, proved to be the most economical process, ultimately supplemented by centrifuges developed in Europe. The two methods now provide the slightly enriched uranium (3 to 5 percent ^{235}U) used in present nuclear power reactors. Because of their construction and type of fuel, reactors cannot explode as does a bomb.

Separation of uranium isotopes by gaseous diffusion. The uranium-235 atoms pass through the "barrier" more readily than the uranium-238 atoms. Thousands of such units are connected.

The World's First Nuclear Reactor

The first nuclear reactor was built at the University of Chicago in 1942 under the leadership of Enrico Fermi. The questions to be answered were, first, whether a controlled chain reaction involving neutrons and uranium could be achieved and, second, whether irradiation of uranium by neutrons could produce enough plutonium to build a bomb.

The reactor shown in the artist's sketch was constructed of chunks of natural uranium as metal and oxide embedded in graphite blocks. The graphite (carbon) served to slow the neutrons to low energy; that is, it served as "moderator." This first uncooled "pile" went critical on December 2, 1942, and eventually reached a power of 200 watts.

Artist's sketch of Fermi's chain-reacting pile at the University of Chicago, 1942.

The success of the Chicago reactor led to the construction of a large plutonium production reactor at the Hanford Engineer Works, located near the present city of Richland, Washington. This reactor also used graphite and natural uranium, with the uranium metal canned in aluminum to protect it from the effects of water coolant. The pieces of uranium, "slugs," were pushed through the reactor after they were irradiated. After irradiation, the slugs were called "spent fuel."

A chemical operation called "reprocessing" was then applied to recover the plutonium that had been produced by neutron irradiation of ^{238}U. The aluminum coating was removed by sodium hydroxide, and the uranium was dissolved by nitric acid. The chemical bismuth phosphate was then added to precipitate the plutonium, leaving uranium and fission products in solution. A separate process was required to remove the uranium.

One disadvantage of the process was that rather large volumes of solids were mixed with the fission products. At the time, of course, the object was to get plutonium, and the waste problem was given a much lower priority. Eventually, a total of eight reactors were built, which operated until 1971. As the result of reprocessing spent fuel, large volumes of the waste were accumulated during and since the war, as the nation stockpiled nuclear weapons for the Cold War with the U.S.S.R. These defense wastes exist and must be dealt with, regardless of the rate of production of wastes from commercial nuclear power.

Many excellent books have been written about the development and use of the first nuclear weapons. The Reference section lists a few of these.

CHAPTER 9

Defense and Development

After the end of World War II in 1945, the United States investigated ways to use nuclear energy for peaceful purposes and, at the same time, sought to maintain and extend the nation's weapons capability. Policies made and actions taken in the period since 1945 have shaped the present state of development of nuclear energy, with its accomplishments, opportunities, and problems.

National Laboratories for Research

To administer research and development on both peaceful and military applications of nuclear energy, Congress created in 1946 the U.S. Atomic Energy Commission (AEC). The Atomic Energy Act of 1946 charged the AEC to provide materials for defense purposes and to carry out weapons tests, to further the successful application of fission for nuclear power, and to find ways to use the new radioactive materials for beneficial purposes.

Several "national laboratories" were designated to carry out related research and development. The first of these were Oak Ridge National Laboratory in Tennessee, Argonne National Laboratory near Chicago, Brookhaven National Laboratory on Long Island, and Los Alamos National Laboratory in New Mexico. Additional centers in later years included Lawrence Livermore National Laboratory and Lawrence Berkeley National Laboratory near San Francisco, Sandia National Laboratories in New Mexico, Idaho National Engineering and Environmental Laboratory near Idaho Falls, the Pacific Northwest National Laboratory near Richland, Washington, and others. For links on the Internet to all the laboratories, see References.

Defense Projects

The national nuclear defense of the United States was supported by several weapons production facilities, administered by the AEC. These included isotope separation plants using gaseous diffusion at Oak Ridge, Tennessee (pictured), at Paducah, Kentucky, and at Portsmouth, Ohio. These facilities have also supplied most of the fuel for research, test, and commercial reactors. For the continued production of plutonium, the Hanford reactors (also pictured) were operated until 1971.

The gaseous diffusion plant K-25 for separating uranium isotopes at Oak Ridge, Tennessee. (Courtesy of Department of Energy.)

Plutonium-producing reactor facilities at Hanford. These were operated for the first time on December 17, 1944. (Courtesy of the Department of Energy.)

The plutonium generated in the AEC reactors is called "weapons-grade" plutonium, containing a rather small concentration of the isotope plutonium-240. This isotope is undesirable because some of its nuclei undergo spontaneous fission, yielding neutrons, which tend to cause premature detonation and inefficient explosion in a weapon. In contrast, "reactor-grade" plutonium, produced by long irradiation, has a large plutonium-240 content and is not useful for constructing a weapon.

For the generation of tritium, used as an ingredient of the thermonuclear weapon (hydrogen bomb or H-bomb), production reactors were built at the Savannah River Plant near Aiken, South Carolina. The exact nature of nuclear weapons and the stockpile of material is not publicly known, but the weapons capability at its peak included some 2000 launchers and 10,000 warheads. In contrast with the energy yield of the World War II bombs of about 20,000 equivalent tons of TNT, the yield of hydrogen weapons is in the megatons. Part of the materials called "defense wastes" come from the chemical treatment (reprocessing) of spent fuel from the production reactors and from the fabrication of plutonium weapons.

A weapons testing program was carried out over the many years of the Cold War, to improve yield and to study military effects. Sites included Pacific bases at Bikini, Eniwetok, and the Marshall Islands, and especially at the Nevada Test Site near Las Vegas. Significant radioactive fallout was experienced, leading to a ban by treaty on aboveground tests.

The Nuclear Navy

The original production reactors were composed of graphite, with cylindrical natural uranium fuel rods. The availability of uranium enriched in ^{235}U made possible a new type of reactor, consisting of metal alloy fuel plates, with water as both cooling agent (coolant) and moderator. This reactor was investigated at the Oak Ridge National Laboratory as a possible power source for naval vessels, especially submarines. Between 1948 and 1953 a submarine reactor was built and tested at Idaho Falls, Idaho, by Argonne National Laboratory and Westinghouse Electric Corporation, under the leadership of Admiral H. G. Rickover.

The submarine *Nautilus* went to sea in 1955 powered by this nuclear reactor using enriched uranium fuel. Because the weight of fuel used was so small, the submarine was able to stay under water for months and to go under the ice at the North Pole. With its first core loading, it went 62,000 miles.

Conventional submarines were replaced with nuclear-powered vessels because of the greater speed, freedom from noise, and ability to remain submerged for a long time. By cutting a reactor-powered submarine in two and inserting a new section, the first guided missile submarine was built. If necessary the rockets could deliver nuclear weapons thousands of miles away. The older Polaris missiles have been replaced with Poseidon missiles, Tomahawk cruise missiles, and Trident ballistic missiles.

Few people realize the size of the active fleet of the U.S. nuclear Navy. It is constantly changing as new ships are built, others are overhauled or modernized, and still others are retired from service. At maximum strength, there were around 150 submarines, 9 cruisers, and 7 aircraft carriers. The *Enterprise* shown in the photograph was the first of its type. It had eight nuclear reactors, 80 aircraft, and a crew of over 5000. For many years, the spent fuel from the naval reactors was reprocessed at Idaho Falls, but since 1992 has been stored there.

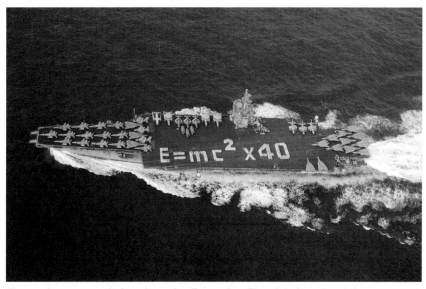

The nuclear-powered aircraft carrier *Enterprise*. Einstein's formula relating mass and energy is spelled out by crew members. The "x40" marks 40 years of U.S. Naval nuclear power. (Courtesy of the U.S. Navy. Photograph by Photographer's Mate 3rd Class Douglass M. Pearlman.)

With the reduction in tensions between the United States and Russia, decisions are being made about the number and character of nuclear naval vessels to be retained. Some new nuclear waste will be developed as ships are decommissioned.

Electrical Generation

The Atomic Energy Act, amended in 1954, called for encouragement of nuclear reactor development for commercial electric power. Industrial organizations cooperated with the AEC in programs designed to learn what reactor types were suitable for economical production of electricity. The homogeneous reactor consisted of a water solution of a uranium salt, with both fuel and moderator circulated through heat removal equipment. Corrosion problems led to abandonment of this concept. A graphite-moderated reactor similar to the production types but cooled with liquid sodium was tested and found impractical. A reactor was tried that was cooled with a petroleum-derived liquid with boiling point much higher than that of water. The cooling agent turned out to be affected adversely by radiation.

Successful, however, was the Experimental Breeder Reactor, at Arco, Idaho (pictured). In 1951 it produced the first commercial nuclear electric power. It was of the fast-reactor type, without moderator, and cooled by liquid metal.

A breeder is a reactor that takes advantage of the number of neutrons, an average of about three, emitted in fission caused by fast neutrons in the element plutonium. Although one neutron must be used to continue the chain reaction, another is available to be absorbed by the fertile isotope uranium-238 to produce additional plutonium. More fuel can be produced than is burned to obtain power.

The advantage of the breeder is that it can use the abundant uranium-238 (converted into plutonium-239) as fuel while other reactors use mainly the scarcer uranium-235. The total uranium resources of the earth are available if the breeder is adopted. Over the years, many countries, including the United States, Russia, United Kingdom, France, and Japan, carried out breeder reactor research and development. For a variety of reasons, including the ready availability of uranium, there is little recent interest in the concept. One promising complete system called the Integral Fast Reactor (see References) was studied at Argonne National Laboratory.

By the mid-1960s two commercial versions of what is called the light-water reactor had been developed. One is the pressurized water reactor, developed by Westinghouse Electric Corporation as a spinoff of the submarine program. The other is the boiling water reactor, first tested at Argonne National Laboratory and perfected by the General Electric Company. In the period 1965-1970, large numbers of reactors were ordered by U.S. utilities, and a major construction program was begun. By 2002 more than 100 power reactors were operating in the United States and three times as many abroad. The photo (p. 58) shows the exterior of a typical nuclear power plant, and the schematic drawing shows the basic system. One other major type of reactor, CANDU, with heavy water, containing deuterium as the

(Below Left) Side view of EBR-I showing the radiation shield

(Above Right) The first use of electric power from atomic energy, December 20, 1951.

The Experimental Breeder Reactor (EBR-I) at Arco, Idaho.
(Courtesy of Idaho Operations Office, Department of Energy.)

moderator, is in operation in Canada, India, and Pakistan. Gas-cooled reactors have been used extensively in the United Kingdom, and the graphite-moderated Chernobyl type is found in the Ukraine.

The H. B. Robinson nuclear plant at Hartsville, South Carolina, operated by Progress Energy. On the far right is the reactor containment building; on the left is the turbine-generator. (Courtesy of Nuclear Energy Institute.)

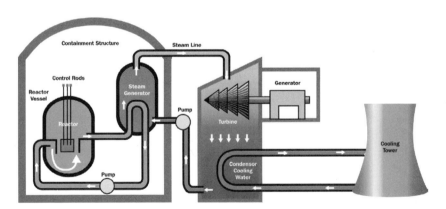

Flow of fluids in the pressurized water reactor (PWR). Water is heated by the nuclear fuel but kept under pressure so it will not boil. It is piped from the pressure vessel to a steam generator. There it transfers its heat to a second supply of water, which boils to make steam for the turbine. (Courtesy of Virtual Nuclear Tourist web site by Joseph Gonyeau—see References.)

A perspective of the present role of nuclear power in the United States can be gained from some statistics. First is the energy produced from different sources, shown in the following table. The energy is used for transportation (27.0 percent), residential (20.7 percent), and commercial and industrial (52.3 percent) purposes. As we see in the other table, the electricity produced by commercial nuclear power plants exceeds that from all other plants except those using coal.

The annual retail sales of electricity (from all sources) in 1999 were 3312 billion kWh at an average of 6.66 cent/kWh. Thus the total sales value of electrical power was $221 billion.

Total Energy Produced in the U.S. by Primary Type
(From Department of Energy, see References)

Energy Source	Energy in Quadrillions of Btu*	Percent
Petroleum	12.4	17.2
Coal	22.7	31.5
Natural Gas and Liquids	22.3	31.1
Nuclear Power	8.0	11.1
Hydroelectric	2.8	3.9
Wood Waste and Alcohol	3.3	4.6
Other	0.4	0.6
Total	71.9	100.0

*British thermal units. One Btu of heat energy will raise the temperature of a pound of water by one degree Fahrenheit.

Electrical Energy Produced in the U.S.
(From Department of Energy, see References)

Type of Plant	Energy in billion kWh	Percent
Coal	1884.3	50.9
Nuclear	728.3	19.7
Natural Gases	569.5	15.4
Hydroelectric	313.2	8.5
Petroleum	123.6	3.3
Wood	37.6	1.0
Waste	23.5	0.6
Geothermal	16.8	0.5
Wind	4.5	0.1
Solar	0.8	-
Total	3702.1	100.0

CHAPTER 10

Uses of Isotopes and Radiation

In accord with the Atomic Energy Act of 1946, the AEC encouraged the development of new applications of radioisotopes and radiation. The objective was to find beneficial uses that would provide direct economic return, advance scientific knowledge, and more generally improve human life. Uses were found in many areas, including medical diagnosis and treatment, agriculture, industry, research, and space travel.

A review of just a few applications of nuclear technology will provide historical background and set the stage for consideration of the management of low-level wastes, which are by-products of the use of isotopes as well as the operation of nuclear power plants.

Tracers

Complicated physical, chemical, and biological processes can be studied by the use of isotopes that "trace" the material of interest. For example, the radioisotope sodium-24, formed by neutron absorption in ordinary sodium-23, has a half-life of 15 hours. If a salt solution containing a small amount of sodium-24 is injected into a person's vein, the speed of blood flow through the body can be measured by the radioactive emanations.

INJECTION OF RADIOACTIVE MATERIAL   DETECTION

FLUID FLOW IN PIPE

THE SPEED OF FLOW OF LIQUID IN A PIPE CAN BE FOUND BY TIMING THE ARRIVAL OF RADIOACTIVITY.

The flow of oil or other fluid in a long pipeline is measured by injecting a small amount of radioisotope at one end and detecting its passage at the other end, as shown in the sketch.

The isotope phosphorus-32, half-life 14.3 days, can be mixed with fertilizer applied to the roots of plants, as sketched here. The radioactive substance is then taken up by the plant,

FERTILIZER

PLANT

GROUND

UPTAKE OF FERTILIZER CAN BE GAUGED BY MEASURING THE ACTIVITY OF PHOSPHORUS-32 IN PLANT LEAVES.

Two uses of radioactive tracers.

and its presence in the stem and leaves reveals where the fertilizer goes. Effects of placement, timing, and amounts of fertilizer can be deduced using this method.

If a radioactive species of an element is used in a chemical compound, the substance is said to be "labeled," i.e., given a special identification. Tritium (hydrogen-3), as an isotope of hydrogen with a half-life of 12.3 years, and radiocarbon (carbon-14), half-life 5715 years, are excellent isotopes for labeling organic compounds (which contain carbon, hydrogen, oxygen, and nitrogen) in a great variety of biological research studies. Many molecules have carbon atoms in different places, so several differently labeled compounds can be formed. Of the isotopes listed below, technetium-99m (6 hours) is the most popular. Tracers of many compounds of technetium can be formed. Its half-life and gamma radiation are favorable for measurements but safe for patients.

Imaging

Nuclear medicine is a recognized specialization that uses nuclear techniques to diagnose ailments. Radioactive materials are used in a process called imaging. The patient is given a solution containing a small concentration of an isotope that has a special affinity for tissue where a difficulty is suspected. A gamma-ray-sensitive radiation detector then scans that part of the body. The pattern of measured radiation reveals information about size, shape, and condition of the organ. The method is widely employed for certain infections, gall bladder and heart diseases, and bone cancer. The table and the sketch indicate how this technique works.

Radioisotopes Used in Imaging of Organs of the Body

Iodine-125, -131	Thyroid, liver, kidney, heart, lung, brain
Chromium-51	Spleen
Technetium-99m	Brain, liver, spleen, kidney, lung, bone
Selenium-75	Pancreas
Strontium-85	Bone

Locating a brain tumor by imaging—injecting a radioactive chemical. The arrow near the top of the head points to the malignancy. (Courtesy of Robert T. Morrison, M.D., Vancouver, British Columbia, General Hospital.)

A special type of nuclear imaging for research goes by the acronym PET (positron emission tomography). One injects a compound containing a positron emitter such as fluorine-18. Combination of the positron with an electron yields two gamma rays. This imaging research shows promise of detecting the onset of Alzheimer's disease.

Therapy

The treatment of disease by the use of radiation from radioisotopes is a common practice. Some of the diagnostic isotopes can be used with higher activity levels to irradiate organs. For example, radiation from phosphorus is applied in leukemia therapy, and that from iodine can treat hyperthyroidism or brain tumors. Several iodine isotopes are available: iodine-125, half-life 59 days, comes from the cyclotron, a particle accelerator; iodine-131, half-life 8 days, is a fission product.

The radiation from cobalt-60, half-life 5.3 years, consists of two gamma rays with an average energy 1.25 MeV. These rays continue to be used widely in the world for the treatment of cancer, even though there is an increasing application of accelerator-produced radiation.

Implantation of a capsule of radioactive material in an organ provides local irradiation. In earlier times, radium was used; currently popular isotopes are iridium-192 (74 days), iodine-125 (59 days), and palladium-103 (17 days). Rice-sized "seeds" are used to treat prostate cancer as an alternative to surgery.

An especially valuable radioactive pharmaceutical for therapy is strontium-89 (50.5 days), which goes to the bone and alleviates intense pain associated with metastatic bone cancer. Other shorter-lived substances being tested in a program of the International Atomic Energy Agency are rhenium-186 (3.8 days) and samarium-153 (1.9 days). Generally, beta or alpha emitters are preferred to gamma emitters because of the greater localization of energy deposition.

Pharmaceutical Research

The development of new drugs for curative purposes by pharmaceutical companies requires extensive use of radioactive materials. Regulations of the U.S. Food and Drug Administration (FDA) call for test data on the behavior of a proposed drug. Tests with animals and humans are required. Tracer studies employ tritium and carbon-14 but also involve a variety of other isotopes, including phosphorus-32 (14 days), chromium-51 (28 days), iodine-125 (59 days), sulfur-35 (87 days), cobalt-57 (272 days), and chlorine-36 (3×10^5 yr), the only long-lived isotope of chlorine. Several of these have half-lives long enough that residues form a waste requiring disposal rather than holding for decay.

When a drug itself is radioactive, data on the absorbed dose to specific organs and the whole body must be supplied to get FDA approval.

Research reactors serve as the principal source of radioactive materials around the world. In the United States, neutron-induced radionuclides are supplied by Canada; local cyclotrons produce charged particle-induced isotopes.

Radiography

X-rays have been used in medical diagnosis for many decades. A refinement in radiography is CT (computed tomography), in which x-ray scattering data are processed by a computer to give very precise information about thin layers of the body. Modern diagnosis involves a combination of nuclear and non-nuclear methods, including nuclear medicine, PET, CT, ultrasonics, and magnetic resonance.

Radiography is also applied in industry for the inspection of metal parts for flaws, as shown in the sketch. Cobalt-60 gamma rays are preferable to x-rays. Advantages are (1) the high energy for good penetration of metal and (2) portability, achieved without the need for an electrical supply, as an x-ray machine requires. The method is especially good for checking welded metal joints in a nuclear reactor cooling system and for locating areas of metal fatigue in aircraft structures.

Gauging

The rays from radioisotopes are useful for making a variety of measurements. One example is the continuous testing of the thickness of paper during its manufacture (as sketched on this page) by the use of beta particles from krypton-85. Variations are detected in the number of particles that get through the paper. In soils the moisture content is measured by the migration of neutrons coming from the reaction of alpha particles on beryllium. Also, as pictured, the density of a fluid moving in a pipe can be found without taking a sample by detecting the gamma rays that get through the pipe. Similarly, the level of liquid in cans of soft drink on a conveyor belt can be checked, at a rate of 2000 cans per minute.

FLAW IN A LARGE METAL CASTING BEING DETECTED BY COBALT GAMMA RAYS

THICKNESS OF PAPER BEING MEASURED DURING MANUFACTURE.

DENSITY OF A LIQUID FLOWING IN A PIPE BEING FOUND BY OBSERVING GAMMA RAY INTENSITY.

Industrial uses of radiation.

Dating

The age of archeological and historical objects is found by the carbon-dating technique. There has always been a certain amount of carbon-14 in the atmosphere. Plants such as trees use carbon dioxide, and traces of this 5715-year half-life isotope are deposited in their tissues. At any later time, measurement of the carbon-14 content tells us the age of any artifact made from the plant.

The age of mineral deposits can be found by examining the ratio of uranium and lead isotopes, ^{238}U and ^{206}Pb. (Recall the uranium radioactive decay chain given on p. 16.) A recently deposited mineral would have little lead; an ancient one would have a great deal. The age of the earth has been estimated by this method.

Neutron Activation Analysis

The absorption of neutrons by many stable isotopes renders them radioactive, and detection of the resulting radiation indicates the amount and type of the original substance. Extremely minute amounts can be measured by this process called "neutron activation analysis." Examples are: industrial mercury pollution in water that is taken up by animals and deposited in their tissues; crime investigation, in which the composition of stolen goods is compared with the original stock; authentication of old paintings by testing the paint for agreement with that available in earlier times; measuring an alloy for minute traces of undesirable elements. An explanation of the extinction of dinosaurs comes from measurement of the iridium content of layers in seabed sediments. It is proposed that a collision of an asteroid with the earth created dust that reduced sunlight needed by plants on which the dinosaurs depended.

Elimination of Pathogens

Radiation is used to kill pathogens (disease-causing agents such as bacteria, fungi, and viruses). Foods can be kept from spoiling by the use of gamma rays from cobalt-60 or cesium-137, without the use of chemicals that may be carcinogenic (cancer-causing). Alternatively, high-speed electrons from a particle accelerator are used (see References). The limited doses applied kill organisms but produce no radioactivity or harmful chemicals. Treatment can suppress trichina in pork and salmonella in chicken and reduce the number of food-borne illnesses and deaths. Food irradiation is practiced in a number of countries around the world, with strong recommendation by the World Health Organization for its use as a safe way to reduce food losses caused by insects and spoilage. Only recently has food irradiation been applied in the United States, even though the FDA has given approval for irradiation of many food products. Spices and strawberries are the principal commodities commercially irradiated.

Medical supplies and pharmaceuticals are sterilized by gamma irradiation. Contamination is prevented by irradiating equipment and materials within sealed plastic containers.

Blood used for a transfusion to a patient undergoing an organ transplant is treated with radiation from radioactive substances to reduce the number of white blood cells and thus prevent attack and rejection of the organ.

Elimination of Insect Pests

Certain insect pests can be controlled by using a special radiation technique. An example is the screwworm fly, whose larvae can kill cattle; it was recently eradicated in the southern part of the United States. Large numbers of males of that species were sterilized by gamma irradiation and then released. Since their mating resulted in no offspring, the screwworm population dropped rapidly. Emergency action in Libya was taken in 1988-1990 upon discovery of a screwworm infestation. A "fly factory" in Mexico supplied millions of sterilized males, which were dropped by plane like crop dusting. Within five months, the pest was eliminated, thus protecting African wildlife. The tsetse fly, a carrier of sleeping sickness, was eradicated in 1997 in the island of Zanzibar. A major campaign of this sterile insect technique (SIT) is under way in a large area of Africa (see References).

Special Power Sources

The heat produced by the beta-decay of strontium-90 is useful for thermoelectric generators, which produce small but steady, reliable amounts of electric power for remote locations. One navigational beacon powered by strontium-90 was in continuous operation for 16 years.

The isotope plutonium-238, half-life 87.7 years, is produced by neutron irradiation in a nuclear reactor. The isotope is an alpha emitter (5.5 MeV) that is used in thermoelectric power generators for space missions. Equipment for the exploration of the moon by *Apollo* and of other planets by *Pioneer, Viking,* and *Voyager* has been powered by such radioisotope devices. *Voyager 2* started with 470 watts of electric power delivered by a radioisotope thermoelectric generator (RTG). It visited Jupiter, Saturn, Uranus, and Neptune, sending photographs back to Earth by radio signals. The RTG power source is necessary at the outer planets because sunlight is too weak for the use of solar panels. *Voyager* is exploring the heliopause, which is the outer extent of the solar wind and magnetic field (see References).

The spacecraft *Galileo*, equipped with a higher power RTG, provided pictures in 1994 of the collision of the Shoemaker-Levy comet fragments with the planet Jupiter.

Voyager spacecraft with radioisotope thermoelectric generator as power source. (Courtesy of the National Aeronautics and Space Administration.) .

Smoke detectors for fire protection use a heavy artificial radioactive isotope, americium-241, half-life 432 years (see References). The detector consists of an ionization chamber with a small radioactive source of 0.7 μCi, sealed in stainless steel. The source continuously emits alpha particles to ionize the air and allow a small current to flow between two plates. Combustion products from a fire enter the chamber and change the current and voltage. An associated electric circuit produces the alarm. This type of smoke detector responds very quickly to a broad range of particle sizes.

Resultant Wastes

Each application of isotopes involves a certain amount of radioactive waste. Some arises in producing the isotopes and in fabricating the devices; some remains after completing an experiment, or a medical test, or at the end of the useful life of a piece of equipment. The degree of hazard resulting from the radioactivity depends on many factors—the amount of material, the rate at which it decays, the energy and type of radiation, whether the substance can enter the human body, and how long it remains in the body. Obviously, an isotope of half-life 15 hours, such as sodium-24, will be essentially gone after a week's storage, at which time the material can be discarded. Others such as 5.3-year cobalt-60 require precautions over a longer time span.* There have been a few examples where a stolen or discarded radiation source was opened and people were exposed to its radioactive contents. The worst of these cases was in Goiania, Brazil (see References).

The foregoing account reveals the ironic fact that the very radiation that can harm human beings can also benefit health and general welfare.

People often ask why we do not use all the radioactive wastes for beneficial purposes and thus reduce the need for storage and disposal. The answer is related to supply and demand, and thus economics. There is an abundance of radioactive materials in spent reactor fuel, but chemical recovery and purification is required to extract the isotopes. The demand for radionuclides as yet does not warrant large-scale processing. The wastes could be a source of heat, but the cost of preparing devices and of providing radiation protection would be prohibitive except for special applications. In summary, there are more radioactive substances than we currently can use economically.

Protection Against Terrorism

National and international actions in response to the attacks of September 11, 2001, are in a state of evolution, as assessment of the dangers and protective measures is made. It is certain, however, that radiation and radioisotopes can be applied to help in counterterrorism. It is not possible here to comprehensively address the problem, but some information may be useful.

X-rays. The use of x-rays in checking the contents of airline luggage at airports is well established. Portable machines using cobalt gamma rays can be helpful in examining shipments from suspect locations.

Detectors. The devices used in the nuclear industry to detect a variety of radiations are available to screen boxes or packages that might contain radioactive material intended as a "dirty bomb." Of special value are those that could detect gamma emitters cobalt-60 or cesium-137.

Weapons. A neutron source with a gamma ray detector can recognize a significant mass of enriched uranium or of plutonium that might be brought to critical (estimated to be 25 kg of uranium and 8 kg of plutonium).

Biological materials. Gamma rays can successfully detoxify anthrax bacteria or other toxic substances in mail envelopes or small packages.

In summary, much of the procedures and equipment required to foil terrorism is available, can be adapted, or is under active development. For additional information, see References.

---

*A good discussion of both the uses of radiation and its effects is found in Eric J. Hall, *Radiation and Life*, 3rd ed., McGraw-Hill, 1993.

CHAPTER 11

Classification of Wastes

Nuclear materials comprise a great variety of isotopes, elements, chemical compounds, and mixtures. Definitions are important because substances are managed by law and regulation according to classification.

Radioactive materials fall into several categories according to their origin, the type of material present, and their level of radioactivity. The first and broadest distinction is:

- Defense
- Commercial.

Defense Wastes

Defense wastes have been generated over the period during and since World War II, at three main Department of Energy (DOE) installations—the Hanford Site near Richland, Washington; Idaho National Engineering and Environmental Laboratory, near Idaho Falls, Idaho; and the Savannah River Plant near Aiken, South Carolina. Plutonium and other isotopes were separated from production reactor spent fuel at Hanford and at Savannah River, while naval propulsion reactor spent fuel was processed in Idaho. In each case the chemical process has left a residue of fission product waste. Other plutonium-contaminated wastes have evolved from weapons fabrication at Rocky Flats, Colorado, and many other sites.

Commercial Wastes

Commercial wastes are those produced by reactors used for electrical power, by facilities used to process reactor fuels, and by a variety of institutions and industries. There is only a small volume of commercial wastes from reprocessing because most of the fuel from power reactors has been left in the form of irradiated fuel assemblies, the spent fuel. Used fuel remains highly radioactive for years after it is removed from the reactor.

The only reprocessing of U.S. commercial wastes was by Nuclear Fuel Services, Incorporated, at West Valley, New York, in the period

1966 to 1972. This plant was shut down because it was uneconomical to operate.

Since 1972 spent fuel has been accumulating at nuclear power plants. As more reactors came on line, the rate of growth increased. The total activity at present greatly exceeds that of the earlier by-products of reprocessing.

Comparisons and contrasts among the principal wastes can be noted. Both defense and commercial reactor wastes stem from the operation of fission reactors, but defense wastes generally are considerably older and less radioactive. Chemical separations used during World War II were designed to extract plutonium well but not to minimize residual wastes. Even though there is a small volume of fission products in defense wastes, the chemicals with which they are mixed add considerable volume. For all practical purposes, there are no separated commercial reactor wastes, but the volume of spent fuel is large because of uranium and metal structures. Institutional and industrial wastes, including those generated by universities and hospitals, generally have a low radioactivity level since they do not usually contain fission products. The volume is rather large, however.

Three Important Types of Wastes

Another distinction among radioactive wastes is:

- High-level
- Transuranic
- Low-level.

High-level wastes (HLW) are those resulting from reprocessing spent fuel or are the spent fuel itself, either of defense or commercial origin. When spent fuel is chemically processed, the residue consists of fission products and small amounts of plutonium isotopes. Such HLW and spent fuel are the candidates for disposal by burial deep in the ground. Geologic disposal is discussed in Chapters 20 and 21.

Transuranic wastes (TRU) are those containing isotopes above uranium in the periodic table of chemical elements. They are the by-products of fuel assembly and weapons fabrication and of reprocessing operations. Their radioactivity level generally is low, but since they contain several long-lived isotopes, they must be managed separately. This classification is composed of isotopes with half-lives greater than 20 years and giving a total activity of greater than 100 nanocuries* per gram of waste material. Isotopes include plutonium-239, half-life 2.411×10^4 years; americium-241, 432.2 years; americium-243, 7.37×10^3 years; and curium-245, 8.5×10^3 years. A very long half-lived isotope is also present—neptunium-237, 2.14×10^6 years.

---

*Recall that one nanocurie is 37 disintegrations per second.

Transuranic wastes give off very little heat, and most of them can be handled by ordinary methods not requiring remote control. For many years they were buried in shallow trenches, but since 1970 they have been placed in retrievable storage. Some plutonium-contaminated soil at Rocky Flats in Colorado, resulting from fires and leaks, is being cleaned up. Disposal of TRU will be discussed in Chapter 20.

Mill tailings are the residue from the physical and chemical processing of uranium ore to obtain uranium. Tailings are only slightly radioactive, but there is an enormous volume of them. This substance, also called "11e(2) byproduct material," is discussed in Chapter 15.

Low-level wastes (LLW) are officially defined as all wastes other than those defined above. The bulk of LLW have relatively little radioactivity and contain practically no transuranic elements. Most of them require little or no shielding, may be handled by direct contact, and may be buried in near-surface facilities. Part of the LLW, however, has high enough radioactivity that they must be given special treatment and disposal. Low-level wastes come from certain reactor operations and from many institutions such as hospitals and research organizations and from industry. Low-level wastes are divided by the U.S. Nuclear Regulatory Commission into three classes—A, B, and C, generally in increasing degree of radioactivity and corresponding need for secure disposal. Those LLW that have even higher activity are called "greater-than-Class-C" wastes, considered to have a hazard comparable to that of spent fuel. Generation and treatment of LLW will be covered in Chapter 16, and disposal in Chapter 19.

Mixed wastes are LLW that contain both hazardous chemicals and radioactive substances. Hazardous wastes are defined as materials that are toxic, corrosive, inflammable, or explosive. They contain specific elements such as lead and mercury, pesticides such as DDT, and cancer-producing compounds such as polychlorinated biphenyls (PCBs). In Chapter 19 the dual regulation of these wastes by the U.S. Environmental Protection Agency and U.S. Nuclear Regulatory Commission is discussed.

There is no standard accepted scheme for listing types of wastes. Other waste classifications are found. For example, in the DOE's environmental impact statement on commercial wastes, one finds these categories, each applied to the word "wastes": primary, secondary, nuclear power plant, spent fuel basin storage, fuel reprocessing plant, mixed-oxide fuel fabrication, and decommissioning. In an official tabulation of waste inventories prepared by Oak Ridge National Laboratory for DOE, one finds these major classes: commercial spent fuel, HLW, TRU, LLW, uranium mill tailings from commercial operations, environmental restoration wastes, commercial decommissioning wastes, and mixed LLW. Part of the wastes now classified as "environmental restoration" were previously identified as "remedial action," related to cleanup actions at DOE sites. The term "legacy wastes" refers to the fact that they were produced during World War II and the Cold War. In Europe, the term intermediate-level waste (ILW) is used for material with radioactivity between HLW and LLW.

Amounts of Defense Wastes

A large volume of wastes classified as high-level defense waste is now stored in underground tanks and bins at three main government sites—Hanford, Idaho, and Savannah River. Estimated amounts at DOE facilities according to type are shown in the next table. About an eighth of the TRU could be retrieved. Not included in the table is a large volume of soil that is slightly contaminated with plutonium.

Waste Volumes in the DOE Complex (in cubic meters)	
High-level waste	339,419
Transuranic waste	108,927
Low-level waste	120,846
Mixed low-level waste	66,480

It is difficult for us to visualize very large quantities such as a million barrels of oil, a billion dollars, and a million cubic feet. To appreciate the total volume figure of 339,419 cubic meters cited in the table, merely note that this is the same as a cube about 229 feet on a side, i.e., sitting on and covering a little over a square acre of land, which is 208.71 feet on a side.

The Hanford plutonium project is seen to have the largest volume of defense wastes. The site was selected immediately after Fermi's successful test of the first nuclear reactor. The semiarid area in southeastern Washington State (see map) was selected because of its remoteness and the availability of water from the Columbia River. During World War II and afterward, the federal government constructed there a total of nine plutonium production reactors, five chemical processing plants, and 177 storage tanks. Of these tanks, 149 were single-walled, constructed between 1943 and 1964; 28 were double-walled, built between 1968 and 1986. Many of the single-walled tanks have leaked as a result of deterioration of the metal with age, plus inadequate attention. Some million gallons are estimated to have gone into the ground. Fortunately, the soil has slowed the movement of some radionuclides into the environment. Hanford researchers continually monitor the movement of the radionuclides in the soil.

High-level defense waste storage at the Hanford Site in Washington State.
(Courtesy of Pacific Northwest National Laboratory.)

In 1990, one of the tanks started to generate hydrogen gas that would vent every 100 days. The tank was given the nickname "Aburp." Concern arose about the possibility of an explosion that would spread radioactivity. To stir the waste and prevent buildup of large amounts of hydrogen, a mixer pump was developed and installed in the tank. Later, chemicals were added to reduce the amount of gas formation. In another set of tanks, the chemical ferrocyanide had been introduced to cause cesium to settle to the bottom. Because ferrocyanide has the potential to ignite or explode when heated above 450°F, concerns were raised about the possibility of the ferrocyanide leading to an explosion inside the tanks. Studies showed that the ferrocyanide had decomposed, significantly reducing the potential for explosion. The removal and permanent disposal of the wastes will take many years.

Throughout World War II and the Cold War, government emphasis had been placed on production of weapons material, and the growing waste problem was neglected. As a consequence, information about the contents of the tanks is inadequate. The investigation of composition and condition must precede remedial action, but it will also generate new waste streams to be managed. Meanwhile, the facilities must be maintained in a safe condition.

In a later chapter, the major challenge of the long-term environmental restoration program for Hanford and the other DOE sites is addressed. This project parallels the other two national tasks— (a) building a suitable repository for HLW and (b) developing sites for the disposal of LLW.

Other Nuclear Materials

Certain other materials requiring physical security, transportation, or disposal are classified and defined as follows:

(a) "Source materials" are uranium or thorium or their ores containing as much as 1/20 percent of those elements. They serve as the source of fissionable material.

(b) "Special nuclear materials" (SNM) are uranium enriched in the isotope uranium-235 or human-made fissile elements such as plutonium or uranium-233. They are special because they can be reactor fuel or weapons.

(c) "By-product material" is radioactive material (excluding SNM) produced by irradiation in reactors or is the residue from the extraction of uranium from ore. It is a by-product from another objective, nuclear power.

CHAPTER 12

Spent Fuel from Nuclear Reactors

The main reactor type in use in the United States and throughout the world is the light-water reactor (LWR). It is so named because it uses ordinary water formed from hydrogen (not deuterium, as in a heavy-water reactor). The water serves as a "moderator," the substance composed of light elements with which neutrons collide and slow down. The water also serves as a "coolant," the medium that removes the fission heat. Two types of LWRs are in use: the pressurized-water reactor (PWR), in which the water is at high pressure and temperature but does not boil, and the boiling-water reactor (BWR), in which steam is produced directly in the reactor by limited boiling at relatively low pressure. Refer again to the coolant flow diagram for the PWR on p. 58.

Reactors in the United States and Abroad

The 107 commercial reactors in the United States are all of the light-water type. Two-thirds are PWRs; one-third are BWRs. As of the end of 2002, 104 power reactors were in operation. Eight others were more than 50 percent completed but startup was indefinite. Some sites have more than one reactor: for example Duke Power Company operates three reactors—Oconee-1, -2, and -3 near Seneca, South Carolina. The total power capacity of the 104 U.S. operating reactors is nearly 100,000 megawatts electric, which is about 20 percent of the nation's electricity production.

The use of nuclear power abroad continues to grow. The following table shows the status of nuclear plants worldwide. As seen in the table, 31 countries have operable reactors, and two others have reactors under construction. The five nations that have the greatest nuclear power capacity in operation are the United States, France, Japan, Russia, and Germany, producing more than two-thirds of the total nuclear power in the world.

Nuclear Fuel Irradiation

Spent fuel has been accumulating at nuclear power plants for a number of years. The older versions of fuel assemblies, described herein, will be the first to be disposed of in a high-level waste

World Nuclear Power Plants, Operating and Total, as of December 31, 2002

Adapted with permission from the March 2003 *Nuclear News*.
(Copyright © 2003 by the American Nuclear Society, Inc.)

Country	Number of Reactors		Megawatts Electrical	
	Operating	Total	Operating	Total
Argentina	2	3	1,018	1,710
Armenia	1	1	376	376
Belgium	7	7	5,680	5,680
Brazil	2	3	1,901	3,176
Bulgaria	4	4	2,722	2,722
Canada	22	22	15,113	15,113
China	7	11	5,426	8,764
Czech Republic	4	6	1,648	3,610
Finland	4	4	2,656	2,656
France	59	59	63,203	63,203
Germany	20	20	22,594	22,594
Hungary	4	4	1,755	1,755
India	14	22	2,548	6,128
Iran	0	1	0	915
Japan	53	58	44,041	48,883
Lithuania	2	2	2,370	2,370
Mexico	2	2	1,364	1,364
Netherlands	1	1	452	452
North Korea	0	2	0	2,000
Pakistan	2	2	425	425
Romania	1	5	655	3,135
Russia	27	33	20,799	26,074
Slovakia	6	8	2,512	3,392
Slovenia	1	1	656	656
South Africa	2	2	1,800	1,800
South Korea	18	22	14,970	18,970
Spain	9	9	7,565	7,565
Sweden	11	11	9,460	9,460
Switzerland	5	5	3,220	3,220
Taiwan	6	8	4,884	7,584
Ukraine	13	18	11,195	15,945
United Kingdom	31	31	11,802	11,802
United States	104	107	99,034	102,637
Non-U.S.	340	387	264,810	303,499
Total	444	494	363,844	406,136

repository. The reactors use uranium that has a higher percentage of ^{235}U (3 percent) than is found in nature (0.7 percent). The fuel is in the form of uranium dioxide, UO_2, as small pellets about 3/16 inch in diameter and 3/8-inch long, as pictured here. These are inserted into 14-foot-long, thin-walled (0.025-inch) tubes composed of an alloy of the element zirconium. Because the metal tube surrounds the fuel, it is often called "cladding"—a coating that prevents radioactive fission products from getting into the cooling water. The tube also provides support for the fuel. After the pellets have been introduced, the ends of the tubes are sealed to prevent water from getting in and gaseous fission products from getting out. Bundles of about 200 of the resulting fuel rods are formed, and the necessary space is maintained between rods as shown in the diagram. These fuel assemblies are about 8 inches on a side, are 14 feet long, and weigh about 1200 pounds, but the assemblies are readily handled with suitable hoists and cranes.

Fuel pellets for a pressurized water reactor. The uranium oxide contains ^{235}U at 3 percent enrichment. (Courtesy of AgipNucleare.)

Nuclear reactor fuel assembly. Bundles of 200 fuel rods, 14 feet in length, 8 inches on a side, are formed into an assembly. (Courtesy of Babcock and Wilcox Company.)

About 180 of the assemblies are closely packed vertically into what is called the reactor core, located in the lower part of the reactor vessel, pictured in the diagram. This vessel has a thick steel wall to withstand the high pressure, 2200 pounds per square inch, resulting from operation with water at 600°F. Cooling water enters the vessel through large pipes welded to the side, comes down around the outside of the core, and is forced up past the fuel tubes, removing the heat of fission and keeping the rods at a reasonable temperature. As fuel is consumed, the reactor is held at a steady power level by adjusting the concentration of the neutron-absorbing boric acid in the cooling water. To change the power of the reactor or to shut the reactor down, the position of special metal control rods is adjusted.

The typical operating period of a reactor is 1 year (extended in recent years, as noted later). At the end of that period, the operators take the top of the reactor vessel off and remove about one-third (60) of the assemblies on the basis of neutron exposure. This highly neutron-irradiated fuel is called "spent fuel" as it cannot sustain a chain reaction. However, it still contains some of the original ^{235}U and most of the ^{238}U. The remaining two-thirds are rearranged to

Reactor vessel of a pressurized water reactor. Cooling water enters the side, goes down the inside wall of the vessel, then comes up through the core containing uranium. (Courtesy of Babcock and Wilcox Company.)

give the best power, and a fresh one-third is inserted. The vessel is then closed and operation begins again.

New Isotopes in Used Fuel

During the 3 years that the fuel has been in the reactor, the irradiation of assemblies by neutrons has consumed some of the uranium and produced some new material. The content of uranium-235 as the main fissile material has been reduced from about 3 percent to about 1 percent, while the uranium-238 has gone down from 97 percent to 94 percent. A new fissile isotope has been produced by a neutron irradiation. Capture of a neutron in uranium-238 followed by beta decay leads to plutonium-239. This isotope complements the uranium-235 as fuel. It has a half-life of around 24 thousand years, emitting an alpha particle of 5.1 MeV energy. Other isotopes of elements above uranium in the periodic table (transuranic elements) are also produced. By successive neutron absorption ^{239}Pu becomes ^{240}Pu, fissile ^{241}Pu, ^{242}Pu, and short-lived ^{243}Pu. The following chart shows the "before" and "after" composition of spent fuel.

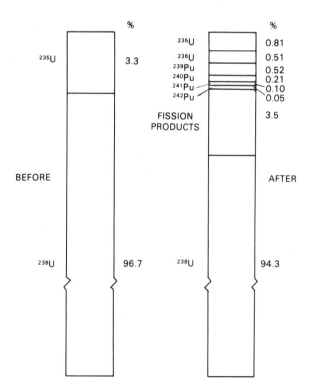

Compositions (in percent) of fresh fuel and spent fuel. Uranium-235 is burned to form fission products and ^{236}U; ^{238}U is converted into plutonium.

The spent fuel still has most of its original ^{238}U and a fairly high fissile fuel content. The total of ^{235}U, ^{239}Pu, and ^{241}Pu is 1.43 percent. The fuel removed is not really as "spent" as it might seem. The light-water reactors are called "converters" because they transform some of the ^{238}U into plutonium isotopes.

Developments

The first large-scale nuclear power plants went into operation in the 1960s. Since then, there have been a number of improvements in performance. The length of the operating cycle, the time between refuelings, has been increased from 1 year to 18 months, which results in an improvement in capacity factor. By increasing the ^{235}U enrichment from the range 2-3 percent to 4-5 percent and changing from a 15x15 array of fuel rods to a 17x17 array, the amount of energy from a PWR fuel assembly has been increased.

A parameter that measures the effectiveness of fuel use is the "burnup," which is the number of thermal megawatt days of energy per metric ton (1000 kg) of fuel. Earlier reactors achieved up to 30,000 MWd/t while later versions reach 50,000 MWd/t. The following table* shows the "before" and "after" compositions in weight percent (w/o) for operation with an initial loading of 4 percent ^{235}U and a burnup of 50,000 MWd/t. In comparison with the figures for the older spent fuel, we note that the newer fuel has less ^{235}U and ^{238}U, and more ^{239}Pu and fission products.

The consumption of fissionable elements during operation of a reactor and the creation of fission products obviously makes spent fuel less able to multiply neutrons than fresh fuel. That fact is taken into account in determining the limit on the number of assemblies that can be put in a shipping container to be assured that criticality will not be achieved. A "burnup credit" is applied that contrasts the neutron multiplication of fresh and spent fuel.

Spent Fuel from Nuclear Reactors

Fresh Fuel (w/o)		Spent Fuel (w/o)	
^{234}U	0.03	^{234}U	0.02
^{235}U	4.00	^{235}U	0.67
^{236}U	0.02	^{236}U	0.56
^{238}U	95.95	^{238}U	92.28
Total	100.00	^{237}Np	0.07
		^{239}Np	0.02
		^{238}Pu	0.03
		^{239}Pu	0.61
		^{240}Pu	0.30
		^{241}Pu	0.18
		^{242}Pu	0.10
		Fission products	5.11
		Other actinides	0.05
		Total	100.00

*Data courtesy of John C. Wagner, Oak Ridge National Laboratory (see References).

CHAPTER 13

Storage of Spent Fuel

Radiation and Heat from Spent Fuel

The spent fuel taken from the reactor after it has operated for a year or so is highly radioactive. The potential radiation dosage at contact with the surface of the fuel assemblies then is millions of rems an hour. Recalling that the lethal dose is around 400 rem, we can see that the fuel must be handled with great care. Moreover, it would be difficult to steal spent fuel without experiencing serious personal risk.

The spent fuel continues to be a source of heat and radiation after removal from the reactor, and thus is stored under water in a deep pool at the reactor site. The water keeps the fuel assemblies cool, and it acts as a shielding material to protect workers from gamma radiation. The water is kept free of minerals that would corrode the fuel tubes.

Fuel Storage Methods

The fuel is no longer suitable for operation in a reactor, but precautions must still be taken to avoid accidental criticality. The assemblies are kept separated in the pool by metal racks that leave about 1 foot between centers. This grid structure is composed of metal containing boron, which absorbs neutrons and prevents their multiplication. The diagrams depict these fuel storage racks, and the photograph shows a spent fuel pool.

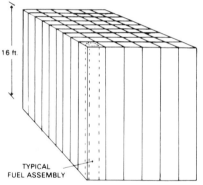

16 ft.

TYPICAL
FUEL ASSEMBLY

PERSPECTIVE VIEW OF METAL RACK, LOCATED IN A WATER POOL.

12.8 in.

TOP VIEW OF THE 6 by 9 ARRAY OF OPENINGS FOR SPENT FUEL.

Spent fuel storage rack to hold 54 fuel assemblies.

81

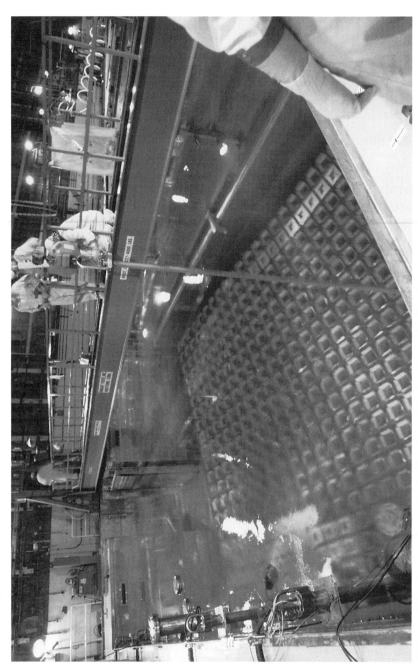

Water pool for storage of spent fuel. (Courtesy of Nuclear Energy Institute.)

The Nuclear Regulatory Commission (NRC) has oversight on the design, construction, and maintenance of spent fuel pools. Regulatory Guide 3.4 on the calculation of heat generation is available on the Web.

Electric utilities that built reactors in the 1960s and early 1970s expected to send their spent fuel to a reprocessing plant within a few months after its removal from the reactor. Indeed, the Barnwell Nuclear Fuels Plant was constructed at Barnwell, South Carolina, for that purpose. There, uranium and plutonium were to be extracted and recycled, with the fission product wastes stored for "cooling" and ultimate disposal.

However, a 1977 federal moratorium on reprocessing was instituted, requiring the utilities to keep the spent fuel at their reactor sites. The requirement to store more fuel than anticipated was met by building closer-packed racks, reducing the spacing between fuel assemblies from 20 inches to 12 inches. Storage in existing pools is only a temporary measure because the spent fuel keeps coming— from 40 to 60 assemblies a year for each operating reactor. As discussed later, the federal government through DOE will take spent fuel for eventual disposal starting around 2010. Meanwhile, however, U.S. utilities are responsible for the fuel. Some companies have transferred fuel from an operating plant to the pool of one under construction. Emergency storage space at national laboratories is quite limited, requiring that other solutions to the storage problem be found. One choice would be to build additional pools at individual reactor sites, at considerable expense because of requirements for a special cooling system, for a water purifier, and for stability against earthquakes. An away-from reactor concept would involve a large storage pool that serves several nuclear plants.

Alternative Storage Methods

To pack more fuel into existing pool space is preferable to constructing new pools. One way to increase capacity is to consolidate fuel. As tested by Duke Power and Westinghouse, fuel assemblies are remotely taken apart and tighter bundles are formed. Twice as much fuel can be put in the same place, and the process can be carried out without damage to the fuel. Consolidation is a satisfactory technique if the pool structure can support the extra weight.

Dry storage of spent fuel has the advantage of avoiding the need for new water pools. Containers can be procured as needed, and little maintenance is required. Virginia Power Company's dry cask demonstration facility at their Surry plant may be a model for many other utilities awaiting permanent disposal capability. The program of design, development, and testing involved many other organizations: a German firm, General Nuclear Systems, designed and built the casks; the Electric Power Research Institute (EPRI) provided financial support; DOE, EG&G Idaho, Bechtel, and Pacific Northwest Laboratory did the testing at Idaho National Engineering Laboratory; and NRC licensed the facility. Design and safety evaluations included radiation levels and the effects of temperature, wind, tornado, fire, lightning, snow and ice, earthquake, and aircraft crash.

Array of spent fuel dry storage casks. (Courtesy of Nuclear Energy Institute.)

The body of the CASTOR V/21 cylindrical cask is cast iron 16 feet high and 7.9 feet in diameter, with walls that are 14.9 inches thick. Fins on the outside help remove decay heat and maintain a safe cladding temperature. The cask will hold 21 pressurized water reactor fuel assemblies in a stainless steel basket to provide proper separation. The total loaded weight is 120 tons.

The empty cask is lowered by special crane into the fuel pool, and assemblies are moved individually from the storage area and inserted in the cask. Double-sealed stainless steel lids are bolted on the top. Water is drained out, and vacuum is applied to dry the fuel completely; then chemically inert helium is introduced under pressure to protect the cladding against corrosion. The cask is lifted out and put on an A-frame on wheels, which is pulled by truck to the storage area, where the cask is unloaded on a concrete pad 3 feet thick, as shown in the photo. Eventually 84 casks on three pads will be installed. The cost of $1.2 million per cask is much lower than the cost of comparable pool storage.

Another design, by Westinghouse, holds 24 assemblies and discharges 12.6 kW of heat with the fuel rod surface temperature no higher than 140°F. The radiation dosage on the outside of the storage containers is 70 mrem/hr. Carolina Power and Light Company has built some large concrete storage modules that accept horizontal fuel casks. A hydraulic system is used to move the casks.

As an alternative to building casks at individual plants, plans are underway for a centralized facility. A consortium of eight companies titled Private Fuel Storage LLC is seeking license to accept and store spent fuel on the reservation of the Skull Valley Band of Goshute Indians in Utah. In 4000 casks around 40,000 metric tons would be stored for up to 20 years plus an extension if needed. The project is viewed as a benefit to all participants (see References).

When DOE starts accepting spent fuel, it will require some storage space. One concept is the Monitored Retrievable Storage (MRS), a large facility located geographically between the generating companies and the fuel disposal site. The fuel would be repackaged at the MRS for disposal. Some devices that could be used for storing spent fuel are (a) an air-cooled, shielded concrete vault, (b) a caisson, consisting of a metal-lined hole in the ground with a concrete plug, and (c) an aboveground concrete cask with natural convection air-flow cooling.

In the wake of the events of September 11, 2001, questions have been raised about the vulnerability of dry storage casks to terrorist actions. In a study by Pennington and McGough (see References), it is reported that if a Boeing 747 aircraft were to strike a concrete cask, the less dense projectile would "flow" around the denser target. No release of radioactivity is predicted, even if there is an accompanying fire.

Weights and Volumes

Let us consider how much material is involved in the management of the older spent fuel that will be the first to be disposed. The assembly has a uranium weight of around 1000 pounds; other metals add some 200 pounds. The volume of one assembly is 6.7 ft^3, so the 60 assemblies removed per year from each reactor give around 400 ft^3 of spent fuel. From 100 reactors in the United States, the annual spent fuel volume would thus be 40,000 ft^3, corresponding to less than 1-foot depth in a standard football field (300 ft by 160 ft). The actual amount of radioactive material in the spent fuel is considerably smaller. From the previous chapter, we saw that fission products have a weight of 3.5 percent of the spent fuel. If the fuel were reprocessed, these would be extracted, as would most of the plutonium isotopes. The uranium would be cleaned up in preparation for reuse. For each 1200-pound assembly, there would be only around 35 pounds of fission product waste. For 60 assemblies discharged per year, this is 2100 pounds, or about one ton. Assuming that the weight of these fission product elements on the average is 10 times that of water (which weighs 62.4 pounds per cubic foot), their actual volume would be only 3.4 ft^3, which is 18 inches on a side. This figure is the source of the claim that a year's wastes from a reactor could fit under an office desk. The statement is misleading because most wastes would not be stored or disposed of in such a concentrated form because its radioactivity would produce intense heat. The wastes would be diluted by some solid before being discarded. This simple sketch compares volumes of fission products and spent fuel produced by one reactor in a year. Choices of what to do with spent fuel are: (a) store it indefinitely in the form as removed from the reactor, (b) reprocess to extract fission products and recycle other materials, or (c) dispose by burial or other isolation technique. In the next chapter, these choices are examined further.

SPENT FUEL
400 CUBIC FEET
PER YEAR

FISSION PRODUCTS
3.4 CUBIC FEET PER YEAR

Comparison of the volume of fission products with the volume of spent fuel. The numbers refer to the production by one reactor in a year.

CHAPTER 14

Reprocessing, Recycling, and Resources

Between the mining of uranium and the final disposal of waste products are many processes; these comprise the nuclear fuel cycle. The term "front end" of the cycle refers to preparation of uranium for use in power reactors; the term "back end" refers to operations performed on spent fuel.

The "Once-Through" Cycle

The sketch outlines the steps in the "once-through" cycle for nuclear reactors as it has been operating in the nuclear industry. First, exploration for new uranium deposits is done. Then uranium is mined in several western states of the United States—mainly Colorado, Utah, and Wyoming. Natural uranium (0.7 percent ^{235}U, 99.3 percent ^{238}U) is extracted from the low-grade ores by a milling process, leaving a large waste residue called mill "tailings."

A complex uranium compound called "yellow cake" is the useful product purified by chemical refining. The compound is then converted to uranium hexafluoride, UF_6, a gas at ordinary temperatures and pressures. UF_6 is shipped to the isotope separation plant, where a process of gaseous diffusion separates an enriched or product stream of ^{235}U content greater than that of natural uranium from a "tails" stream of depleted uranium. The slightly enriched uranium (3 to 5 percent ^{235}U), still as gaseous

EXPLORATION

MINING   MILLING

PROCESSING

ENRICHING

FUEL FABRICATION

REACTOR

INTERIM STORAGE

The nuclear fuel cycle, carrying uranium from the mine to spent fuel storage in water pools.

87

UF$_6$, goes to the fuel fabrication plant, where it is converted to the form of solid pellets of uranium dioxide, UO$_2$. The pellets are inserted into tubes to form fuel rods, which are sealed, tested, and assembled into bundles for shipment to the reactor site. Each fuel assembly remains in the converter reactor for about 3 years, after which it is removed and placed in a storage pool for radioactive "cooling."

There are now two choices of what to do with the spent fuel. The first is to continue interim storage in water-filled pools in a "once-through" cycle. The fuel assemblies would then be stored or disposed of intact or in bundles of fuel rods in a waste repository. The second choice is to recycle uranium and plutonium. This is possible through reprocessing the spent fuel by methods similar to those used at Hanford during World War II. Plutonium can be blended with slightly enriched uranium to form a fuel called "mixed oxide" (MOX), which is a combination of UO$_2$ and plutonium dioxide, PuO$_2$. The uranium in spent fuel has a ^{235}U content higher than that of natural uranium. It can be returned to the isotope separation plant for reenrichment and reuse. The diagram shows the once-through and recycle modes of operating the nuclear fuel cycle, including points at which wastes arise.

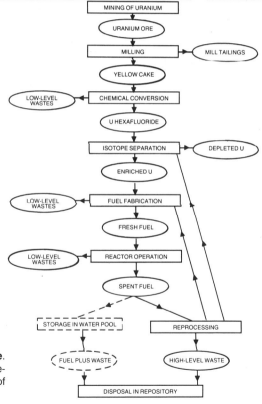

Wastes in the nuclear fuel cycle. Two choices are shown: the once-through cycle and a recycle of uranium and plutonium.

Reprocessing of Spent Fuel

The long-range plans of the nuclear industry of the 1970s included reprocessing spent fuel.* Thus, when the activity in the fuel stored in the water pool had dropped sufficiently, the fuel would be placed in casks for shipment to a centralized reprocessing plant. Some commercial reprocessing was done by Nuclear Fuel Services at West Valley, New York. A large-scale reprocessing facility designated as the Barnwell Nuclear Fuels Plant was built in South Carolina by Allied Gulf Nuclear Services (AGNS). It was tested with nonradioactive chemicals. There it was planned to cut fuel assemblies into pieces, dissolve the contents of the fuel tubes with nitric acid, and apply a plutonium-uranium solvent extraction (PUREX) method. This would separate the three principal components: uranium, plutonium, and fission products. The flow of materials is seen in the diagram. This expectation was not realized. The AGNS facility was never used for radioactive waste and was ultimately dismantled.

Reprocessing was banned as a U.S. national policy by President Carter in 1977 because of the belief that it makes plutonium more readily accessible. It was hoped that countries not yet having nuclear weapons would thus be discouraged from developing reprocessing facilities to obtain plutonium. President Reagan in 1981 rescinded the ban and urged the nuclear industry to resume commercial reprocessing. Financial problems and uncertainty about future government policy, however, have discouraged industry from acting.

Other countries however have successfully implemented reprocessing. A facility at Sellafield, United Kingdom, is operated by British Nuclear Fuels LTD (see References). The plant processes spent fuel from the graphite Magnox reactors of the country, spent fuel from PWRs, and used MOX fuel assemblies from Japan. The fuel is first stored for at least 5 years to allow decay of fission products, after which it is transferred to a feed pond. A shearing machine chops fuel rods into small pieces, exposing the uranium oxide. These segments are introduced to a vat of hot nitric acid, which dissolves the fuel, leaving the "hulls" of metal. These are sent to a separate plant for treatment and disposal. A centrifuge removes solids from the solution, containing uranium, plutonium, and fission products. Pulse columns use air and a series of perforated plates to enhance the separation of organic and aqueous phases. The separated uranium and plutonium solutions are concentrated by evaporation and converted into powders by spraying them into a hot vessel. The

---

*For an excellent discussion of reprocessing, read the article by William P. Bebbington, "The Reprocessing of Nuclear Fuels," *Scientific American*, December 1976, p. 30. Highly technical details on reprocessing are found in the book by Justin T. Long, *Engineering for Nuclear Fuel Reprocessing*, American Nuclear Society, 1978.

powders are then packed in stainless steel containers for storage or shipment. Plutonium can be returned to the source or mixed with natural uranium to produce MOX. Fission product waste is sent to storage.

A similar process is carried out by the other major reprocessor, the company COGEMA, which operates a plant at La Hague in France (see References). There, two plants, UP3 and UP2 800, have a capacity of 1700 metric tons per year. They use an advanced PUREX process with special features: a continuous nitric acid dissolver for the uranium oxide; non-corrosive zirconium vessels; and a large-capacity unit to mix fission products with powdered glass and melt them together for eventual disposal in solid form. Two other COGEMA plants produce a total of 195 tonnes of MOX per year, mainly for the French nuclear program EDF, for Japan, and for Germany.

France strongly favors reprocessing, for several stated reasons:

(a) The process is ecologically conscious, involving recycling of 96 percent of the input spent fuel.

(b) Plutonium is put in the best form for transport to the country of origin.

(c) The process is economical, being a small fraction of the cost of producing nuclear power.

(d) It yields a very small volume of radioactive waste to dispose of.

(e) It is a money-making project, "comparable to the share France would receive from the sale of 1000 Airbus airliners . . ."

(f) Success in reprocessing demonstrates global technological leadership.

To that list one can add the fact that by recycling uranium and plutonium the amount of mining and milling of new uranium ore is reduced and uranium reserves are conserved. The removal of plutonium and its subsequent use as a component of fresh fuel eliminates it as a source of hazard in waste disposal. Of course, stringent controls of the plutonium must be maintained to avoid diversion to unstable countries or to terrorists.

Breeder Reactor

The fissile content of spent fuel is equivalent to uranium enriched to about 1.5 percent in ^{235}U, as noted in Chapter 12. This is twice the fissile content of natural uranium. More than half the atoms are plutonium, which can be used as starting fuel for "breeder" reactors. The breeder is a reactor that does not have a neutron moderator, and thus fission is initiated by fast neutrons. When fueled by plutonium with the core surrounded by ^{238}U as a "blanket," the breeder produces more fissile material than it burns. In 10 to 20 years, it can create a new core of plutonium. Recall from Chapter 12 that the converter reactor depends on ^{235}U and produces some plutonium. In the once-through mode, it makes use of only about half of the ^{235}U in natural uranium, and very little of the ^{238}U. In contrast, the breeder reactor is designed to make use of the more abundant isotope ^{238}U. The available uranium supply would last around 50 times longer if the breeder cycle were adopted.

Research and development on the liquid metal fast breeder reactor (LMFBR) was carried out starting in the 1950s in several countries. Leaders have been the United States, the Soviet Union, France, the United Kingdom, and Japan. The United States started construction but later abandoned a demonstration breeder called the Clinch River Breeder Reactor Project. The French sodium-cooled fast breeder SuperPhenix was operated for a while and then was shut down because of leaks, probably permanently. As of 2002, there were only two LMFBRs in operation in the world—one in Russia and one in France. The European Fast Reactor is a breeder being designed as a cooperative effort of France, Germany, and Great Britain. It will be several years before the design is completed. Japan continues to favor reprocessing and breeder reactors and may be the first to successfully revive the breeding process, with emphasis on the prototype reactor MONJU (see References).

Decisions about the breeder affect the management of nuclear wastes. Adoption of breeders automatically requires reprocessing; disposal involves fission products rather than spent fuel assemblies.

The desirability of pursuing research, development, and testing of breeder reactors is a subject of debate. Some believe that the breeder is undesirable because of the production of plutonium, which is both toxic and usable for nuclear weapons. Others, however, note that the safest place to put plutonium is in a reactor, which is itself a very strong source of radiation. As breeder reactors are built and go into operation, they may be fueled initially with fissile plutonium from light-water reactors. Alternatively, they could be fueled with plutonium removed from weapons dismantled as a consequence of disarmament treaties. For fertile material, that serving as the source of new plutonium, breeder reactors can use the large stockpile of depleted uranium from uranium isotope separation over many years. This has less ^{235}U (about 0.2 percent) than natural uranium (0.7 percent), but only the ^{238}U is needed to produce new plutonium in the breeder.

Breeder reactors are likely to remain in the research stage around the world for a number of years. On the other hand, breeders may become popular in several foreign countries that do not have large reserves of either fossil fuels or uranium. The breeder's ultimate economic advantage is felt only when fossil fuels are in short supply and when natural uranium stockpiles and reserves become too expensive to exploit. Currently, uranium fuel prices are low and form a rather small part of the cost of nuclear electricity.

Partitioning and Transmutation

A process called "partitioning" separates chemicals as an extension of reprocessing. It extracts, isolates, and irradiates certain long-lived transuranic elements such as neptunium, americium, and curium, and fission products such as iodine-129 (1.7×10^6 yr) and technetium-99 (2.13×10^5 yr). Partitioning is being considered to improve the safety of disposal of high-level waste. In a process called "transmutation," some of the nuclides would be fissioned to become fission products with generally shorter half-lives than the parent nuclei. Others would become stable or have shorter half-lives. The phrase "actinide burning" (actinides are elements above 89 in the periodic table) is often used interchangeably with "transmutation."

Two different alternative systems are available for the treatment of waste. The first consists of a metal fuel fabrication unit at the site, a reactor operating with fast neutrons, and a chemical extraction process based on melting. The Integral Fast Reactor (IFR) concept would combine these three steps. Neutrons in the reactor would bombard and transmute the transuranic and fission product isotopes in the recycled fuel. The IFR project was cancelled by order of the U.S. Congress, but the technology remains promising, especially in light of concerns about the security of plutonium.

The second approach would involve a linear accelerator to bring protons up to an energy of 1600 MeV. A current of 0.1 ampere would bombard a target of a heavy element such as lead or mercury. In that target, a high-energy nuclear process called "spallation" releases as many as 50 fast neutrons per proton. Surrounding the target is a "blanket" containing the separated isotopes, some of which are fissionable. Neutrons are multiplied by factors of 10 to 20 in this subcritical layer, and are absorbed to transmute isotopes into less hazardous forms. The virtue of transmutation is that the volume and activity of waste is greatly reduced, eliminating the need for extensive disposal facilities. In principle, one proton can destroy several hundred transuranic atoms. Thus a few accelerators could process the wastes of all the reactors in the United States, assuming that the spent fuel is reprocessed. The devices would probably not be used to produce electric power because of the inability to ensure continuous accelerator ion beams. The systems would be very safe because they are subcritical; amounts of wastes from fission are considered to not be excessive (see References).

Although these concepts look promising, several technical, economic, and societal issues need to be resolved, as discussed in the references for this chapter.

We now have a picture of the present nuclear fuel cycle and possible variations on it involving recycling of uranium or plutonium. As was seen in the diagram, some radioactive wastes are produced in every stage. The disposal of these in an environmentally sound way will be discussed in subsequent chapters.

CHAPTER 15

Uranium Mill Tailings

The amount of uranium in ore is quite small, about 0.2 percent by weight in the United States. Thus there is a large residue called "tailings" from the chemical processing of ore, called "milling." Each 1000-megawatt electric nuclear reactor requires about 150 tons per year of natural uranium. The total ore required is 500 times this figure. The map shows uranium production centers in the United States, including uranium reserve areas, mills, and other plants. Shortly after the end of World War II, the Atomic Energy Commission encouraged domestic exploration and mining. The product was used for weapons and peaceful applications. However, the demand for nuclear fuel did not grow as expected and foreign supplies increased. Uranium production in the United States declined, and no conventional mills are in operation. Employment (in person-years) declined from 21,950 in 1979 to less than a thousand by 2000. Originally, all

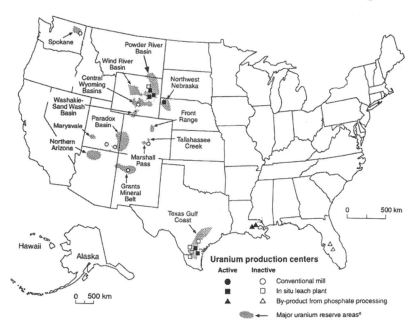

Uranium reserves, mills, and plants in the United States (from "Red Book" of OECD/NEA and IAEA, see References).

uranium ore was extracted by underground mining or by open-pit mining. Later methods include in situ leaching, in which a solution is piped into a deposit to dissolve the uranium, or extraction as a by-product from phosphate mining. Reserves and natural uranium stocks in the United States are ample, augmented by imports of former weapons material from Russia. The product of mills is uranium oxide, U_3O_8, called "yellow cake." Over the years, it has been converted into uranium hexafluoride, UF_6, for isotope separation or to metal for irradiation to produce plutonium. The report *Linking Legacies* (see References) gives the locations, organizations, and processes involved in the production of uranium for weapons and reactors. For many countries around the world, the status of uranium exploration, mining, and importation is described in the "Red Book" (see References).

The tailings contain elements such as thorium and radium, which are mainly by-products of the decay of ^{238}U as was shown in the table on p. 16. Most of these elements are not removed in the extraction of uranium. Generally, tailings leave the mill as a liquid sludge and are allowed to dry. They are collected in piles within enclosures. Precaution is necessary to prevent tailings from contaminating groundwater or getting into the air as dust. For their radioactive content, tailings are no more dangerous than the ore that was removed from the earth, but as the material has been brought to the surface and been converted to a new form, it can pose a hazard unless protective layers of earth are applied.

Uranium mining at McClean Lake, Canada. (Courtesy of Nuclear Energy Institute.)

Radioactivity in Tailings

The tailings are classified as radioactive waste separately from the low-level wastes from power production and from isotope uses. On a short-term basis, the radiation hazard to the general public in the surrounding region from tailings is relatively small. The possibility of erosion by wind and water plus geologic effects requires, however, that the residues be properly stabilized.

One isotope of main concern is radon-222. As we saw on p. 16, it is the daughter of radium-226, which has a half-life of 1599 years.

Radon is produced continuously by the decay of radium left with the ore residue. Radon-222 has a short half-life, 3.82 days, and decays in a chain with four short-lived products. The element radon is an inert gas of the same type as helium and neon. Thus much of it can come out through the pores of the tailings and the soil into the atmosphere and be carried some distance by wind before it decays. The daughters of radon can deposit on surfaces as contamination or can deposit in lung tissue of animals and human beings.

Over the years some 140 million tons of tailings have accumulated. Earlier there was little concern about the potential hazard. Because almost all earth and rock contain uranium and thus radium, the tailings were treated essentially as just another form of ore residue. Regulations set by the states were not very rigorous and their enforcement was often lax. Concern about mill tailings arose as the result of use of the sand-like material for fill in building construction in Grand Junction, Colorado. In 1966 it was realized that occupants were being exposed. Mill owners were requested to take better care of tailings. Congress in 1972 authorized funds for cleanup in Colorado, and in 1978 passed the Uranium Mill Tailings Control Act, calling for the Department of Energy to take remedial action at inactive milling sites. A DOE program involving 24 sites in ten states is in progress.

The passage of the National Environmental Policy Act (NEPA) resulted in a new look at the situation and prompted the Nuclear Regulatory Commission to initiate a generic environmental impact statement on mill tailings. The document has some specific guidelines about ways to handle tailings. It recommends that tailings be located away from population centers, preferably underground, with a cover 3 meters thick to keep the radon radioactivity that escapes below 2 picocuries per square meter per second. It also recommends that the bottom and walls of a tailings pit be covered with a thick layer of clay and that vegetation be grown above the pit to resist erosion. Extensive studies of ways to control migration of the radioactive material have been conducted. Emphasis has been placed on preventing flooding erosion of slopes by the use of riprap (rock armoring).

Some studies show that the potential hazard to the public from mill tailings can be larger than that from other places in the nuclear fuel cycle, including reactors. The exact degree of hazard from radon and its daughters is difficult to specify, however, because there seems to be little or no correlation between lung cancer incidence and geographical areas having a high level of radon. The release of radon is not unique to uranium mining as the mining of phosphates for fertilizers brings up radon in comparable amounts. Much additional information and insight on mill tailings is found in the book by Eisenbud (see References).

Indoor Radon

The radiation exposure from radon was quoted in Chapter 5 as 55 percent of the total exposure for the average American. This figure is not the same for all people because some homes and buildings contain more radon than others. Concern about excessive amounts has been growing recently. As early as 1906, the famous scientist Lord Rayleigh is said to have known of radioactivity in granite, and geologists over the years had studied the uranium content in rocks and water. Attention was called to the problem of radon in homes by Henry Hurwitz, Jr., who noted that federal standards for emissions from nuclear power plants were much more stringent than those on radon. Radiation measurements made in the aftermath of Three Mile Island Unit 2 accident led to the discovery of high radon concentrations in Reading Prong, a geological formation mainly in Pennsylvania. There some homes had concentrations hundreds of times the average.

There is some uncertainty as to the actual hazard, which is estimated in part from cancer cases among uranium miners. These workers have historically been at risk, especially as they tend to be heavy smokers as well. It is difficult to separate the effects of radon and smoking. Also the work environment of miners is quite different from that of a typical person. Another factor is the difficulty in modeling the dose-effect relationship.

Although the hazard is commonly associated with radon, it is the short-lived daughter products—isotopes of polonium, bismuth, and lead—that are the principal offenders. One might think the dose would be due to the radioactivity of decay products arising from inhaled radon. In fact, it is largely the result of breathing air that contains radioactive daughters attached to dust particles along with unattached decay products. The amount of radiation received thus depends on the relative amounts of such materials the body traps or eliminates. Further discussion of the modeling of radon effects, including the use of the working level (WL) unit of exposure, is found in the book by Shapiro (see References).

For many years the contribution of radon to the average public radiation exposure was ignored, but in 1987 the National Council of Radiation Protection and Measurements included it in published estimates. The Environmental Protection Agency mounted a public information program, issuing in 1986 and updating in 1992 the bulletin *A Citizen's Guide to Radon*, which is also available on the Web (see References). The booklet explains the origin and nature of radon and recommends testing of individual homes and corrective action if the radon concentration in air is above 4 picocuries per liter (pCi/l). Using the linear no-threshold hypothesis, EPA estimates that around 14,000 lung cancers appear in the United States each year from radon. This figure would be considerably lower if there is a hormesis effect as discussed in Chapter 5. The EPA emphasizes the added risk of developing lung cancer if one is a smoker. Putting the risk on an individual basis, at the 4 pCi/l level a non-smoker has an increased risk according to EPA of about 1 chance in 1000 of developing lung cancer, while a smoker has 1 chance in 100.

Radon in buildings comes primarily from the ground, which naturally contains a varying amount of uranium. The amount of radon present in the air at any time depends on the degree of ventilation because outside air has a low concentration of around 0.1 pCi/l. The sealing of rooms, however, to conserve energy does not account for the large amounts of radon found in some locations. It appears that the radon is "pumped" through cracks in basement floors or walls by differences in pressure, temperature, and wind velocity.

Methods of control include thorough sealing of cracks and the installation of pipes and fans that will exhaust the gas from under floors to the atmosphere and prevent its getting into the main living areas. The expense of such modifications varies but is estimated to average around $1200 per building. To determine whether action is needed, the EPA recommends that one or perhaps two short-term measurements be made, using commercially available kits costing around $25. Because radon measurement results can fluctuate greatly from time to time, a test lasting several months is in order before incurring a large expense. It is considered better to include radon protection in the original construction of a house than to treat it later. Some researchers (see References) believe that the EPA should make a concerted effort to find and correct levels in excess of 20 pCi/l around the country, thus protecting those who are in greatest hazard.

Appreciation is extended to Dr. James E. Watson, Jr., for his thoughts on the radon problem.

added risk of developing lung cancer if one is a smoker. Putting the risk on an individual basis, at the 4 pCi/l level a non-smoker has an increased risk according to EPA of about 1 chance in 1000 of developing lung cancer, while a smoker has 1 chance in 100.

Radon in buildings comes primarily from the ground, which naturally contains a varying amount of uranium. The amount of radon present in the air at any time depends on the degree of ventilation because outside air has a low concentration of around 0.1 pCi/l. The sealing of rooms, however, to conserve energy does not account for the large amounts of radon found in some locations. It appears that the radon is "pumped" through cracks in basement floors or walls by differences in pressure, temperature, and wind velocity.

Methods of control include thorough sealing of cracks and the installation of pipes and fans that will exhaust the gas from under floors to the atmosphere and prevent its getting into the main living areas. The expense of such modifications varies but is estimated to average around $1200 per building. To determine whether action is needed, the EPA recommends that one or perhaps two short-term measurements be made, using commercially available kits costing around $25. Because radon measurement results can fluctuate greatly from time to time, a test lasting several months is in order before incurring a large expense. It is considered better to include radon protection in the original construction of a house than to treat it later. Some researchers (see References) believe that the EPA should make a concerted effort to find and correct levels in excess of 20 pCi/l around the country, thus protecting those who are in greatest hazard.

Appreciation is extended to Dr. James E. Watson, Jr., for his thoughts on the radon problem.

CHAPTER 16

Generation and Treatment
of Low-Level Waste

How Low-Level Wastes Are Defined

Radioactive wastes are classed officially as low-level wastes (LLW) if they are not spent fuel, high-level reprocessed wastes, mill tailings, or transuranic wastes. In general, LLW contain a small amount of radioactive material in a rather large volume. Thus they usually do not require shielding or heat-removing equipment, and most are acceptable for near-surface disposal, i.e., above, on, or slightly below ground level. They are mainly by-products of the operation of nuclear reactors and the use of radioisotopes. Although some LLW come from DOE and defense programs, this chapter focuses on commercial wastes, both those from the nuclear fuel cycle and those from the "non-fuel-cycle," which include industrial and institutional wastes. This chapter will discuss also two special types of LLW, coming from decontamination and decommissioning.

Low-Level Wastes from Nuclear Power Plants

Of the low-level radioactive wastes generated today, more than half of the volume and most of the activity comes from nuclear power plants. The radioactivity arises from two nuclear processes: activation and fission.

Activation results from nuclei absorbing neutrons in the nuclear reactor. Targets are metals in the core and vessel, corrosion products in the cooling water or on surfaces, or impurities in the water. Because much of the internal structure of a reactor is composed of stainless steel, isotopes of iron, cobalt, nickel, and manganese are produced. An example is cobalt-60, half-life 5.27 years, emitting gamma rays of 1.25 MeV average. It is formed by a slow neutron reaction with cobalt-59. The table lists data on the important activation products. A certain amount of tritium (hydrogen-3), with half-life 12.3 years, is present as a result of occasional fission of uranium into three particles. Tritium also is found in the cooling water of reactors that use boron as a soluble control absorber.

Fission products serve as the second large source of low-level radioactivity. Although most of these are retained within the fuel tubes, a few appear in the cooling water. Some are from neutron

Activation Products in Reactor Coolant

Isotope	Half-life, yr	Radiation Emitted	Parent Isotope
Carbon-14	5715	Beta	Nitrogen-14*
Iron-55	2.73	x-ray	Iron-54
Cobalt-60	5.27	Beta, gamma	Cobalt-59
Nickel-59	7.6×10^4	x-ray	Nickel-58
Nickel-63	100	Beta	Nickel-62
Niobium-94	2.4×10^4	Beta, gamma	Niobium-93
Technetium-99	2.13×10^5	Beta	Molybdenum-98
			Molybdenum-99[†]

*(n,p) reaction
†Beta decay

irradiation of small amounts of uranium left on the outside of fuel rods during fabrication or residual impurity of natural uranium. Additional fission products come from pinhole corrosion leaks in the fuel tubes. Because it is impractical to shut the reactor down to remove an offending fuel assembly, some leakage into the water is permitted. Some of the most important beta-gamma-emitting fission products are iodine-131 (8 days), cerium-144 (285 days), cesium-134 (2.1 yr), cesium-137 (30.2 yr), technetium-99 (2.13×10^5 yr), and iodine-129 (1.7×10^7 yr). Many other chemical species, however, are represented in the coolant, including carbon-14 (5715 yr) resulting from neutron bombardment of nitrogen in air.

There are various choices of management of waste materials from a nuclear plant: (a) send to a conventional landfill if it can be assured that the radioactivity is negligible; (b) process with the plant's own waste treatment capability; (c) send to a commercial processing facility; (d) send directly to a low-level waste disposal site; or (e) store at the plant. Among the treatment options are filtration, evaporation, incineration, compaction, and solidification. In the next section, these processes are described.

Volume Reduction Techniques

The cooling systems of reactors are closed loops, but leaks of water occur, and radioactive components must be repaired or replaced, resulting in the spread of contamination. Also, the coolant is treated to remove impurities, giving rise in time to significant volumes of radioactive material. Because of increasing costs of shipping and disposal of low-level wastes, nuclear utilities have expanded their onsite processing facilities.

Waste management programs include preventing contamination, segregating wastes by type, reducing volume, and conditioning of materials in preparation for shipment and disposal. Many of the practices are dictated by compliance with federal regulations on waste disposal. The principal methods, applied to low-level wastes both at nuclear power plants and defense waste sites, are as follows.

Filtration is much like the process of straining grounds in a coffee pot. The mixture of solid particles and water is forced through a medium that is porous to liquid, as shown in the diagram. When the filter cartridges accumulate enough material, they are removed and packed in a drum for disposal. The 55-gallon steel drum is very popular because of its low cost and convenient size.

Filter for removing solid particles from water. One of many filter cartridges is sketched. (Adapted from report NUREG/CR-0142 by A. H. Kibbey and H. W. Godbee, Oak Ridge National Laboratory.)

Ion exchange is used if the waste is in solution and cannot be filtered. The process removes contaminants of both the activation and fission types. Charged atoms in the liquid replace other nonradioactive atoms on the solid surface of a resin bed. Eventually the resin becomes saturated and must be either washed clean and regenerated for future use by treatment with a chemical or disposed of as waste.

Evaporation is simply boiling to remove water as steam, leaving the waste more concentrated. A residue of sludge collects in the bottom of the vessel shown in the diagram. The condensate water may be recycled for other uses while the sludge is removed periodically.

Evaporator for reducing waste volume. The product is a concentrated sludge removed from the bottom of the vessel. (Adapted from report NUREG/CR-0142 by A. H. Kibbey and H. W. Godbee, Oak Ridge National Laboratory.)

Incineration is used for a variety of combustible materials. Ordinary materials such as paper, cloth, wood, and plastic become slightly contaminated and can be burned, reducing volume by a factor of 20 or more. The incinerator shown is of Japanese make. The air containing the smoke from burning passes through filters to remove particles. The gases coming from the process thus contain very little radioactivity and can be released safely with dilution to the atmosphere. The ashes are radioactive and must be treated as low-level waste.

Incinerator for combustible materials. Radioactivity concentrates in the ashes, and the gaseous effluent is diluted to safe levels. (Courtesy of Hitachi Zosen Corp. of Japan.)

Compaction is used to reduce the volume of waste and also to produce a more stable structure for disposal. Different degrees of compaction are possible, depending on the force exerted by the compactor and the nature of the material. The small compactor shown has a plate driven by a hydraulic press that can crush the contents of a 55-gallon drum or the drum and contents. The largest heavy-duty machine is Duratek's UltraCompactor™, which can apply a force of ten million pounds to drums or boxes of waste. The device is used to compact dry active waste by factors of six, and can compact other materials such a soils, motors, pipes, and valves.

Compactor for waste drums. The plate can either compress the contents of a 55-gallon drum or crush a drum full of waste. (Courtesy of Consolidated Baling Machine Co., Brooklyn, New York.)

Solidification means mixing wastes with some solid material that will resist attack by water after disposal. One popular material is concrete, in which the water remaining in the low-level waste is taken up by the concrete as it sets. The diagram (on the next page) shows a portable cement supply and a concrete mixer used at a nuclear power plant. Another medium is asphalt. The diagram (on the next page) shows how reactor wastes and asphalt are fed into a heated grinder that pulverizes, removes water, and mixes. The viscous product pours into a 55-gallon drum, where it solidifies. Another material is polyethlyene, which is easy to handle, is relatively light, and has favorable resistance to water. Solidification stabilizes the waste physically and chemically, making it less subject to leaching when the material is buried.

Solidification using concrete. Water needed to produce concrete comes from the wastes. (Courtesy of LN Technologies, Columbia, South Carolina.)

Solidification using asphalt. The heated grinder dries waste, and the molten waste-asphalt mixture solidifies in 55-gallon drums. (Courtesy of Wemer-Pfleiderer Corp.)

Another way to protect wastes from chemical attack is to use an officially designated "high-integrity container" made of plastic or concrete and reinforced for stability. It is approved by the Nuclear Regulatory Commission (NRC) as able to last for 300 years under typical exposure to radiation from the contents.

The choice of volume reduction process and equipment depends on the types of wastes and their volumes, on economic factors, and on regulatory requirements. For example, an incinerator is of no use in handling failed contaminated metal equipment; a compactor will not compress sludges; a million-dollar supercompactor is not needed to handle a few cubic meters per year.

A number of nuclear power plants have installed full waste treatment systems designed to achieve optimum technical and economic performance. A typical example is the Surry Radwaste Facility of Virginia Power. The system collects waste liquid, separates oil and solids, and evaporates water using heat recovery. Concentrated waste is thoroughly mixed with bitumen and solidified for disposal. Effluent water has minimum radioactivity. The facility also has a shredder-compactor and a hot machine shop.

Low-activity fuel cycle wastes also come from the uranium chemical conversion process and from fuel fabrication plants. The volume of these is less than one-tenth that from the nuclear power plants that they supply.

Other Sources of Low-Level Wastes

Non-fuel-cycle wastes are often called "industrial and institutional wastes." They are generated by industrial organizations that supply or use radioactive materials, e.g., pharmaceutical companies, and by institutions such as clinics, hospitals, medical research laboratories, and universities. Medical wastes include animal carcasses and other biological waste, trash, various liquids, sealed radiation sources, and technetium-99m generators. Industrial and institutional wastes stem from many uses of many different radionuclides. Larger generators use some of the same treatment methods employed by nuclear power plants.

One special type of liquid waste is scintillation detector fluid used as tracers in biomedical research and medical tests, and for measurements of radiation. Millions of vials of these solutions, used for counting radioactivity, are disposed of each year. They typically contain the chemicals toluene or xylene, with a slight content of tritium or carbon-14. The beta decay of these isotopes triggers the release of light that is detected by a sensor. The total annual activity in scintillation vials in the United States is small, only about 10 curies per year, but because of their chemicals, they often pose a disposal problem. In some cases the fluids have a low enough activity that they can be disposed of in a sanitary sewer. Disposal sites will not accept such liquids for burial. Incineration is regarded as the best disposal method.

The radioisotope most widely used as a tracer for medical diagnosis is technetium-99m, half-life 6 hours. This gamma emitter is extracted from a longer-lived isotope molybdenum-99, half-life 66 hours. The ^{99m}Tc is said to be "milked" from the ^{99}Mo "cow." The half-life of technetium is so short that holding any residues or contaminated material for decay is preferable to shipment to a disposal site. Reduction by a factor of more than a million occurs in 20 half-lives, which is only 5 days for this isotope. Rather than dispose of the generator (as the ^{99}Mo source is called) when it has weakened, it is preferable to send it back to the supplier, who will combine it with material to produce a new generator.

Several medical isotopes have half-lives of only a few days; these can be held for decay and thus pose no disposal problem. Some institutions have tended, however, to ship such wastes away for disposal to avoid the need for monitoring and surveillance. Only iodine-125 (59.4 days) has too long a half-life for convenient long-term storage. In bioresearch, however, the most important and widely used isotopes are tritium (12.3 yr) and carbon-14 (5715 yr), which must be disposed of by other means.

The management of low-level wastes in a university research laboratory is a challenge because of the variety of wastes of different half-lives and the lack of space for storage (see References). Concern about access to LLW disposal has been expressed by a committee of the National Research Council (see References). The committee addressed the needs of institutions that conduct biomedical research and hospitals where radionuclides are crucial for diagnosis and treatment of disease. The principal finding had to do with escalating costs, which have caused the use of radioisotopes to decline in favor of less-effective nonradioactive alternatives. Costs impact the fraction of funds available for research and add a burden to patients. The problem would be aggravated if new technical developments required increased use of radioactive materials, including longer-lived isotopes, which would require expanded storage facilities or higher disposal costs. Also, any arbitrary closure of disposal facilities would be severely disruptive to research programs and medical applications.

Although it is generally assumed that all radioactivity should be avoided and controlled, there are amounts so low that they can safely be ignored. These are treated as exempt by the NRC in the *Code of Federal Regulations*, Parts 30.15 and 20.2005. Examples are 25 mCi of tritium and 100 μCi of promethium-147 as used in self-luminous watches. Also scintillation detector fluids or animal carcasses may be disposed of as ordinary nonradioactive wastes if the tritium or carbon-14 content is less than 0.05 Ci per gram, with certain limits on annual disposal. Such amounts are termed "*de minimis*"; this comes from the Latin phrase "*de minimis non curat lex*" and is translated as "the law does not concern itself with trifles."

Amounts of Wastes Produced

Knowledge of the characteristics of wastes is important at every stage of management for several reasons—to ensure radiation protection during processing, to achieve proper segregation and packaging, and to meet regulations on shipment and burial. The features that must be known are volume (ft^3 or m^3) according to waste stream, radioactivity as a specific activity (Ci/ft^3 or Ci/m^3) preferably by isotope, and heat generation rate (W/ft^3 or W/m^3). "Source terms" represent such data, which can be used to plan waste management programs.

Wastes shipped from generators to processing or disposal facilities are always accompanied by a manifest, which is a shipping document that lists properties of waste container contents. The generator is responsible for the accuracy of the information, and the disposer makes spot checks to verify. Some of the waste radionuclides have extremely long half-lives and correspondingly small activities, and are difficult to detect. Examples are technetium-99 (2.13 x 10^5 yr) and iodine-129 (1.7 x 10^7 yr). Predictions of amounts are made using "scaling factors" against readily measured fission products like 30-year cesium-137. Detailed studies of actual amounts of the long-lived nuclides in waste are made to establish these factors. This practice assumes that waste streams are rather consistent.

The volumes and activities of LLW from a typical nuclear power plant are given in the table. The reactor is Fermi 2 of Detroit Edison Co., located at Newport, Michigan. It is a General Electric boiling water reactor with 1139-MWe capacity.

Solid Low-Level Wastes for the Three-Year Period 1999–2001 Sent From Fermi 2 to Barnwell

(Data supplied by Earl M. Page.)

Type	Volume, m^3	Activity, Ci	Specific Activity, Ci/m^3
Dry compressible waste	157.9	120.6	0.764
Spent resins	140.8	590.5	4.19
Irradiated components	9.12	9155	1004

The dry compressible waste has the largest volume but a very low specific activity, while the irradiated components are small in volume but very high in activity. In the 1980s, the nuclear utilities greatly reduced the amount of low-level waste by several techniques—avoiding contamination of inert materials, screening to separate radioactive and nonradioactive trash, reducing water leakage, compacting dry active waste, and incineration. The bar graph shows the dramatic decrease per U.S. power reactor over a span of 20 years. The purpose of such efforts was to cut costs of waste transportation and burial.

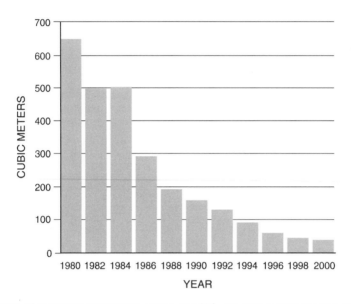

Volume of low-level radioactive waste per U.S. nuclear power reactor, 1980–2000. Boiling water reactors and pressurized water reactors are represented. The number of cubic meters dropped by a factor of more than 17. (Courtesy of the Institute of Nuclear Power Operations and the Nuclear Energy Institute.)

Decontamination

The operation over a period of time of any nuclear installation, be it fuel fabrication plant, nuclear reactor station, or reprocessing facility, involves contamination by radioactive substances. Cleanup during operation is called "decontamination," while steps taken after shutdown are called "decommissioning."

Decontamination means the removal of radioactive material from surfaces such as building floors or walls, hand tools, and the insides of vessels. The object is to reduce radiation exposure to persons working in the area. The technique should be nondestructive and safe to apply; otherwise, personnel might receive more person-rems than the cleanup would save.

Methods are chosen according to the items to be decontaminated. Small tools and equipment can be hand-scrubbed with cleaning agent and cloth or steel wool by an operator wearing a respirator. Other equipment responds to ultrasonic cleaning, in which a solution is vibrated at a high frequency such as 25,000 cycles per second. Removable reactor components may be cleaned also with acid baths, electrical currents, sandblasting, strong chemicals, water jets, and steam. Some surfaces have strippable coatings. Some inaccessible pipes or vessels may be cleaned by circulating high-temperature chemicals.

Large volumes of water can be decontaminated using filters and ion exchangers. The Three Mile Island Unit 2 reactor recovery program provided a major test of decontamination techniques. The program also developed an array of ingenious underwater remotely operated tools to dislodge and safely remove highly radioactive deposits.

The decontamination of the Barnwell Nuclear Fuels Plant, discussed in Chapter 14, was effected efficiently and safely by virtue of excellent planning and coordination. For an account of the special techniques used, see References.

In the future, an increased use of robots will enhance both safety and economy. Some of these electrical/mechanical devices can rapidly sample and analyze the contents of a tank of waste. Others can enter an area, survey it for radiation, and clean up contamination. It is possible to adapt for nuclear use robots that are being developed for military activities, or handling powerful explosives, or hazardous waste operations.

Decommissioning

Decommissioning means the removal of a nuclear power plant or other nuclear facility from service after its useful life and taking the necessary steps to protect the public from residual radiation hazard. The useful life of a reactor is the time between its startup and the point when repairs and replacements are excessively expensive or safety would be compromised by continued operation. Most commercial reactors are licensed to operate for 40 years. In view of the high capital cost of new reactors, strong efforts are made to continue operation for at least that long.

The first action in decommissioning any reactor is to remove and dispose of the spent fuel, which automatically reduces the inventory of radioactive materials greatly. Several choices then are available: (a) "mothballing," officially called SAFSTOR, in which some decontamination is done but then the plant is closed and guarded, perhaps indefinitely, (b) entombment, or ENTOMB, in which concrete and steel barriers are erected to seal in radioactive components, (c) dismantlement, or DECON, involving removal and disposal of all radioactive material. A variant is delayed dismantlement, allowing for decay of many of the radioisotopes. Another possibility is to convert to a non-nuclear heat source, retaining as much of the equipment as possible.

Decommissioning requires considerable planning to carry out the task safely and economically. After fuel is removed, the system will be decontaminated to a certain extent. Next will be disassembly using torches and jackhammers while rigorously controlling the spread of contamination. The internal structures of the reactor vessel will contain activation products of the kinds listed earlier. They will be in the form of deposits on surfaces and embedded in the metal. Cobalt-60 dominates for the first 50 years or so, and over the long term, nickel-59 (7.6×10^4 yr) and niobium-94 (2.4×10^4 yr) are of concern. Many components will be cut up and buried as low-level waste. Highly radioactive materials will exceed limits for near-surface burial and will have to be dealt with separately.

The disposal of metal structures such as pressure vessels and heat exchangers poses special problems in transportation and disposal because of their size. An alternative is melting, which eliminates most of the residual radioactivity and allows recycling of useful material. The 20-ton, 7200-kW melter of Duratek converts contaminated decommissioning metals into useful forms; other products of recycling are canisters for high-level waste and reinforcement bars for concrete. However, release of decontaminated metal for general industrial use is strongly opposed by the steel industry (see References).

Cost estimates for decommissioning vary greatly, but a typical figure is $150 million. This is a relatively small fraction of a $2 billion cost of a plant and is very small compared with the $12 billion value of the electricity produced over the life of the plant.

For additional information on decommissioning, a comprehensive report by the Department of Energy is recommended (see References). There, the main emphasis is on DOE facilities no longer needed.

Projections

The amounts of low-level radioactive waste to be disposed of in the future depend on several factors:

(a) The extent of volume reduction by generators. The utility industry can still effect some further reduction of volume and activity, but large additional changes are unlikely.

(b) The role played by waste processors through recycling programs. Commercial organizations have the capability to decontaminate equipment to a point where the metals could be reused if authorization were achieved.

(c) The use of new technology. Biological wastes with very slight radioactivity can be treated chemically to permit disposal of the residue in a sewer.

(d) The number of reactors that are shut down and decommissioned, in comparison with the number that achieve license extension.

(e) The rate at which new advanced light water reactors or other advanced reactors (e.g., high-temperature gas reactors) are placed in operation.

(f) The possible expansion of radioactive methods for medical diagnosis and therapy to obtain improved health care.

(g) The potential widespread application of nuclear technologies for protection against terrorism.

The factors above in turn depend on local and general economic conditions, social acceptance, and national policy, all of which are difficult to predict even in the short term. It is certain, however, that the technology of safe disposal is available if there is the necessary political will to utilize it.

The results of a study by the Department of Energy of the U.S. nuclear capacity in the past and projection into the future are shown in the graph. The three cases differ in the assumptions made on plant retirements. If new plants came on line around 2010 the nuclear capacity could be maintained constant.

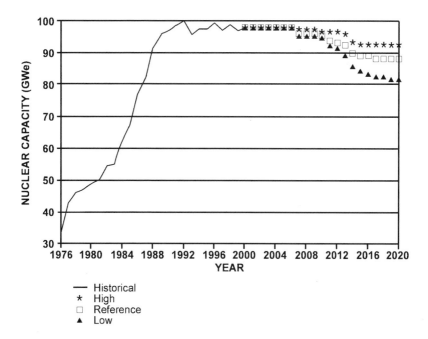

Historical and predicted trends in U.S. nuclear capacity, 1976–2020. Adapted from DOE and NRC reports (see References).

CHAPTER 17

Transportation of Radioactive Materials

Public and private carriers transport radioactive materials of many types all over the country. Many of the shipments are medical isotopes, some are institutional wastes, and many are low-level wastes from nuclear reactor operations. Spent fuel is occasionally transferred between reactor stations. When the federal government begins to accept spent fuel for disposal as high-level waste, the number of shipments will increase significantly.

Concerns About Shipping

People tend to be apprehensive about the transport of radioactive materials through or near their communities. Their concerns stem from a basic fear of radiation, awareness of the likelihood of accidents, and inadequate knowledge of the physical protection provided for such materials. People perceive a hazard from radiation emanating from contents of the vehicle as well as from contamination that might result from rupture of a waste container. One hears one or more statements such as (a) no shipments should be made, (b) shipments should be made to avoid our town, (c) we should know in advance the route to be taken and the schedule of shipment, (d) the vehicles should have a prominent "radioactive" marking, (e) wastes should be shipped inconspicuously to avoid hijacking, (f) guards or an escort vehicle should accompany the shipments, (g) a radiological rescue team should be available at all times, and (h) shippers and carriers should be heavily insured.

Many expectations of the public are being met through physical security and adherence to regulations. However, a complete ban on transport is unrealistic because radioactive materials, including wastes, exist and must be moved. It is not possible to transport wastes without coming near areas with some population by the very nature of the ground transportation system. This chapter gives some accident data and examines the design requirements on packaging of radioactive materials to withstand impact, fire, and immersion.

Facts About Transportation

Some statistics are available on transportation in general and nuclear materials in particular. Some 500 billion packages of commodities of all kinds are transported in the United States each year. Motor carriers handle 57 percent, railroads 38 percent, and waterways about 5 percent. Less than 1 percent goes by air.

Hazardous materials shipments are a very small part (1 in 5000) of all goods and materials transported, and radioactive materials shipments are a rather small part (1 in 50) of the hazardous materials. Therefore, one shipment in 250,000 contains radioactive material. The sketch shows the percentages of shipments, and the tables give some U.S. data on radioactive material transport according to its origin.

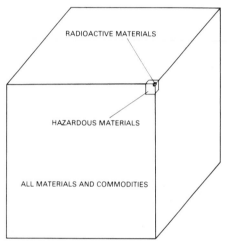

Shipments of goods in the United States. Radioactive materials form 1/50 of hazardous materials, which are 1/5000 of all items shipped.

The level of public safety in the transport of radioactive materials can be put in perspective by examining the table. It refers to one year (2001) in which a breakdown of data by class of hazardous materials is available. A negligible fraction of the incident reports involve radioactive materials. Note, the figures refer only to accidents that are reported to the Department of Transportation. Some events of each type fail to be reported.

The small number of incidents involving radioactivity is borne out by a study conducted by the International Atomic Energy Agency in 1986.* It revealed that fewer data are available than desired but stated regarding transportation, "Reviews of the available historical data have shown that there has never been a serious incident involving the dispersal of radioactive material."

---

*IAEA-TECDOC-398, Vienna, 1986

One Year's Hazardous Materials Incident Reports	
(Source: U.S. Department of Transportation's Hazardous Materials Information System. See References.)	
Class	**Incidents**
Corrosive material	7049
Flammable—combustible liquid	6835
Poisonous materials	1147
Miscellaneous hazardous material from fabrication	588
Oxidizer	456
Nonflammable compressed gas	424
Combustible liquid	302
Flammable gas	204
Organic peroxide	178
Infectious substance	139
Flammable solid	119
Other regulated material, Class D	48
Poisonous gas	39
Spontaneously combustible	16
Dangerous when wet material	14
Explosive no blast hazard	8
Radioactive material	**8**
Explosive mass explosion hazard	3
Explosive fire hazard	2
Very insensitive explosive	1
Total for 2001	17,580

Annual Shipments of Nuclear Materials in the United States	
(Data courtesy of Sandia National Laboratories.)	
Type of Material	**Shipments per Year**
Exempt amount or limited radioactive level, e.g., smoke detectors, luminous signs or watches	700,000
Pharmaceutical and other medical sources, e.g., radioisotopes used for diagnosis or treatment	910,000
Industrial radiation sources, e.g., gages to measure thickness of paper, portable x-ray devices	220,000
Nuclear materials, e.g., uranium, fresh fuel from fabrication plants, interplant spent fuel	200,000
Wastes from all industrial and medical sources besides nuclear power plants	100,000
Nuclear power plant wastes	50,000
Total	2,180,000

From time to time, communities or states pass ordinances or laws prohibiting the movement of radioactive materials. Some of these are not enforced. Confrontation between local and federal jurisdiction is sometimes avoided by rerouting. In other cases, the federal government, through the Department of Transportation, asserts its authority on the basis that national laws take precedence over local laws. The so-called "commerce clause" stating that restraint of trade is illegal is sometimes invoked.

Boxes, Cans, and Casks

The philosophy of protecting the public against radiation in transportation has been that accidents will occur and that protection must be provided by the packaging of the radioactive materials. The degree of protection is selected on the basis of the level and type of radioactivity contained. Type A packages (see the diagram of a typical one) can contain a limited number of curies and are designed to withstand normal wear and tear of transport exclusive of accidents. An inner glass bottle is protected by a metal can, a fiberboard insert, a lead container, and a fiberboard box. Even if the package were badly damaged, there would be little hazard because of the small radioactive content. Type B packages (two views of an example container are shown) are designed for a larger number of curies and must withstand accidents without leaking. Important features are the holddown devices, heavy steel wall, fire protection, and internal suspension.

S FIBERBOARD INSERT
T FIBERBOARD BOX
U TIN CAN
V ABSORBENT PAPER WADDING
W SCOTCH TAPE SEAL
X TOP SECTION LEAD CONTAINER
Y BOTTOM SECTION LEAD CONTAINER

Diagram of a representative type A shipping container, which can vary markedly in size and materials. (Courtesy of Department of Transportation.)

Type B transport cask used by Chem-Nuclear Systems, Inc. The capacity is 4 ft^3, lead shielding is over 10 inches thick; the cask lid weighs 1600 pounds. (Adapted from a Department of Transportation report.)

The shipping cask for spent nuclear fuel is a still more rugged and elaborate container. It provides four types of physical protection:

(1) containment to prevent material from being released into the environment

(2) shielding to prevent radioactive exposure to employees or passersby

(3) heat management to remove the energy released by decaying fission products

(4) criticality prevention to avoid accumulation of enough fissile material to be multiplying.

In the era when reprocessing was anticipated, it was planned that spent fuel would be sent from the nuclear plant to the reprocessing facility after a brief "cooling" period in the water pool. Casks were thus designed with circulating water to remove much of the decay heat. The IF-300 model by the General Electric Company is an example of the early design. The two cutaway views show its features. It will hold 7 PWR or 18 BWR assemblies.

Cutaway view of a spent fuel shipping cask. PWR or BWR fuel assemblies are held in a fuel basket. Water and lead serve as shielding. Fins help remove heat to the air. (Courtesy of Pacific Northwest National Laboratory.)

Spent fuel cask in normal arrangement for shipment by railroad. (Courtesy of General Electric Company.)

More modern versions of spent fuel casks do not contain water because the fuel has been stored for a long time and decay has reduced its heat-generation rate. Some consideration has been given to the use of multi-purpose casks that could be used for storage, transportation, and disposal. Cost and efficiency would be enhanced, but the requirements for each phase are not necessarily the same.

Cask for storing and shipping spent fuel. (Courtesy of EG&G Idaho, Inc.)

Fuel casks are built with massive amounts of fracture-resistant metal such as ductile cast iron or stainless steel. For added protection, large end fittings called impact limiters are attached. These are made of balsa wood or cork, which will absorb much of the energy in any impact.

Until recently, there were relatively few casks available in the United States. In anticipation of eventual acceptance by DOE of spent fuel from nuclear plants, several companies in the United States and abroad have designed casks capable of being licensed by the Nuclear Regulatory Commission.

Such casks, like all of those containing enriched uranium or plutonium, must be safe from nuclear criticality. As in spent fuel pools, protection is provided by separation and use of neutron absorbers. A combination of experiments and computations provides the design features for spent fuel casks. Computer codes such as ORIGEN determine the concentrations of fuel and absorber isotopes, while statistical (Monte Carlo) codes such as KENO V in SCALE are used to estimate neutron multiplication factors. Such information permits the designer to apply burnup credit, noted in Chapter 12 as accounting for the effect of fuel exposure to neutrons in a reactor. Typically a drop in multiplication factor of 20 percent occurs in fuel that has been exposed to 20,000 megawatt-days per metric ton (1000 kg).

For transporting transuranic wastes (TRU) to the Department of Energy's Waste Isolation Pilot Plant (WIPP) in New Mexico (see Chapter 20), a B-type container called TRUPACT II has been built. It is designed for use with "contact-handled" TRU, which is 98 percent of the total. The cask is cylindrical, with a flat bottom and domed lid. It holds two layers of seven 55-gallon drums of waste. The containers, with a capacity of 3290 kilograms, are doubly contained and sealed. They can be carried by truck or rail. TRUPACT II passed the required NRC accident tests and has been in regular use for several years without accident.

Transfer of plutonium from France to Japan by freighter brought protests from people concerned about a possible attack of the vessel, or hijacking by terrorists, or an accident that would contaminate the sea.

Transportation Safety

Wastes are but a small part of the radioactive materials transported, but the same regulations apply to all. On a world basis, the applicable document is the International Atomic Energy Agency's *Regulations for the Safe Transport of Radioactive Material*, first issued in 1959 and frequently updated. These rules are compatible with those of the United States appearing in the Nuclear Regulatory Commission's 10 CFR 71 and the Department of Transportation's 49 CFR 171 and following sections. Such rules derive from recommendations by standards-setting bodies such as International Commission on Radiological Protection (see Chapter 6), which apply fundamental radiological protection principles.

Regulations are based on concepts such as annual limit of intake (Chapter 6). Numerical values are prescribed for the maximum contents of packages in both curies and curies per gram, according to type of container, the material form (concentrated or dispersed), and the particular isotopes involved. For example, the maximum activity for a sealed californium-252 source is 2 Ci; that for normal tritium is 1000 Ci; that for uranium is unlimited, but the number of curies per gram allowed depends on the enrichment.

For many years, shipping casks have been designed to meet stringent rules of the Nuclear Regulatory Commission. A cask must be able to withstand a sequence of events as damaging or more damaging than those that would actually be encountered. The first test is a 30-foot fall on a flat, unyielding surface. Such a test resembles a fall from an overpass onto concrete. However, it should be noted that all real surfaces, such as ground or highways, are somewhat yielding, so that the design requirement corresponds to a drop from a much greater height. The second test is a 40-inch fall onto a metal pin 6 inches in diameter. This is similar to a sharp corner of a bridge abutment. Third is a 30-minute exposure to a fire at a temperature of 1475°F. This is more demanding than a fire from a ruptured tank of gasoline. Finally, the container must not leak when immersed in water for 8 hours. This corresponds to an accident in which a truck rolls off into a creek near the road.

A series of tests was conducted by Sandia National Laboratories to verify the ability of spent fuel shipping casks to meet NRC requirements:

- A tractor-trailer rig carrying a cask was crashed at a speed of 60 mi/hr into a concrete block weighing 690 tons. The photo shows the extensive damage to the truck, with the cask intact

- A cask rail car was crashed into the concrete block at 81 mi/hr

- A locomotive going 80 mi/hr collided with a flatbed truck carrying a cask

- A rail car with cask was exposed for 90 minutes to a fire at 1800°F.

Photograph taken a fraction of a second after collision of a tractor-trailer with a concrete wall at 60 mi/hr. (Courtesy of Sandia National Laboratories.)

Color photographs of the above tests are found on the Sandia Web site (see References).

None of the fuel casks were damaged enough to release radioactivity, and the damage that did occur confirmed the predictions from design analysis and scale-model tests. Subsequent studies include even more unusual situations. Examples are heat from a torch of burning propane from an adjacent ruptured tank car, and crushing barrels by impact with each other within a truck that stops suddenly. Rules do not call for every cask to be tested, but a typical cask of a large number must be. Test facilities in accord with IAEA rules for radioactive transport containers are available in 12 countries.

Supplementing the physical tests are computer simulations, which input features of the casks, subject the model cask to a variety of accident conditions, and determine whether there is damage and release of radioactivity. A description of the computer analysis is found at the Sandia Web site (see References).

Most radioactive materials are carried on highways, with far fewer by air. They include industrial radiation gages, radiography sources and cameras, low-level waste, and spent fuel. Accidents involving radioactive materials over a 24-year period (1971-1994) amounted to only 0.8 percent of all transportation-related accidents. This record can possibly be attributed to superior driver training and motivation. Over that span of time, 1636 events involving radioactivity were reported, including transport accidents, handling accidents, and reported incidents. The one serious accident with A-type containers involved a truck carrying unirradiated fuel that was struck, caught fire, and burned. The most important statistics deal with transport of higher levels of radioactivity using B-type containers. Of the events, only 54 were of that type, and there was no release of radioactivity in any of them.

Transportation of Spent Fuel

When the repository at Yucca Mountain is ready, the Department of Energy will begin shipping spent fuel from the pools or dry storage containers at nuclear power plants. DOE estimates that starting around 2010 there will be 175 shipments per year, mainly by rail. This is considered a small number compared with the 300 million annual shipments of hazardous materials in general. The shipments are expected to continue for 24 years. An overview of spent fuel transportation is found on a DOE Web site (see References).

Concern has been expressed about the possibility of radiation exposure along the routes. It is the responsibility of the Nuclear Regulatory Commission to provide rules and to monitor compliance, to assure the public of the safety of transportation. NRC has made a series of studies over the years to confirm its regulations. In 1977, it issued a "baseline" report (see References), followed in 1987 with a computer-based update called the "modal study" (see References). There it was found that hazards from accidents would be less than previously thought. In 2000 a risk assessment study using probabilistic safety analysis (see References) extended earlier work. Included is the use of a sophisticated computer program RADTRAN, which takes account of the radiation dose to everyone involved—the truck drivers, people in passing cars, persons at truck stops, and those living along the route. The program distinguishes between the accident-free normal operation and postulated accident situations. The calculated doses were minuscule, being fractions of a millirem. The final comment in the summary of the 2000 study was, "Thus, NRC concludes that spent fuel shipments as analyzed are safe to continue." Such a conclusion is vigorously denied by the state of Nevada, which opposes the establishment of the Yucca Mountain repositor: (see References).

A discussion of the methodology for analyzing spent fuel transportation risk is found in a Sandia report, and the RADTRAN program can be accessed on the Internet and exercised (see References).

Preparedness for Emergencies

Despite all precautions, there is some risk in transporting radioactive materials, including wastes. An emergency response capability is available in the United States to cope with waste transportation accidents. The first responders to an accident on the highway would be police or firefighters. They would set up an exclusion area, attempt to save lives, and report to headquarters, starting the notification of the state emergency center. The second responder would be a state radiological protection team who would help in stabilizing the situation. In a serious emergency, governmental assistance can be quickly mobilized. The Federal Emergency Management Agency has a Federal Radiological Emergency Response Plan (FRERP) in concert with 16 other government agencies. They have plans to help in case of a transportation accident as well as one at a nuclear plant. See References for the operational plan of FRERP. The third responder would be a commercial organization that specializes in cleaning up radioactive materials.

CHAPTER 18

Health, Safety, and Environmental Protection

The goal of safe waste disposal is to ensure that practically no radioactive material reaches human beings. Thus in managing radionuclides, account must be taken of the various ways material could be released, transferred, and absorbed. These are called "pathways to man."

Pathways

The artist's sketch shows how the dispersal of radioactivity by air, water, and land is related to the food chain of plants and animals.

Major pathways by which environmentally dispersed radionuclides can affect living organisms. (Courtesy of Pacific Northwest National Laboratory.)

The example shown is an underground source of radioactivity. If erosion by air and water removed soil and exposed the wastes, some particles could enter the air, be deposited on the ground elsewhere, and become a source of direct radiation. The material might also be deposited on plants or get into the soil in which crops are grown. Milk or meat from cattle that graze on the vegetation would then be a pathway to humans. If waste containers leaked into surface water or groundwater, nearby streams or wells could become contaminated. Water used for irrigation could have the same effect on plants as particles from the air. Drinking the water or eating aquatic foods from the stream could also be harmful.

An applicant for a license to operate a disposal facility for radioactive waste must provide evidence that pathways will not result in an excessive dose to a worker or a member of the public.

Multiple Barriers

A "systems approach" has been adopted in the design and construction of disposal facilities for radioactive wastes, whether high-level, low-level, or transuranic. That is, several obstacles are placed between the wastes and habitations. In this "defense in depth" method, the first barrier to movement of contaminants is the waste's physical and chemical form. For low-level wastes, the materials will be dried, incinerated, or compressed. They may then be mixed with a solid such as cement, asphalt, or plastic, in a noncorrosive condition.

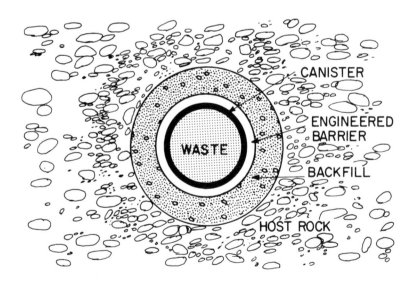

Multiple barriers to prevent waste migration. (Courtesy of Battelle.)

For high-level wastes from reprocessing, the fission products may be melted with glass, as discussed in Chapter 20, or mixed with natural compounds that strongly bind the waste. For spent fuel, the uranium oxide fuel form itself is resistant to corrosion. The rods in assemblies may be consolidated to reduce the volume occupied. The second barrier is the container, which will be composed of a suitable metal, plastic, or ceramic. Corrosion-resistant metals include stainless steel, pure copper, titanium, and various alloys. More than one container may be used. Third is a fill material such as bentonite clay, which swells when it becomes moist and thus prevents the passage of water. Other physical barriers such as concrete containers or walls may also be employed. Finally, the layer that provides the most protection is the geologic medium—the soil or rock—that filters and delays particles of radioactive material from the flowing water.

Each barrier has a role in preventing migration of radioactive material. The waste form and container hold the wastes for several hundred years, during which most of the radionuclides decay to safe levels. Eventually, the waste is dissolved by underground water and may leave the container. The surfaces within the rock, such as particles, pores, grain boundaries, and fissures, all have the effect of retarding the motion of waste material. Most of the chemical species remain within the medium. Because considerable distance is maintained between the disposal site and civilization, any wastes that remain will be highly diluted.

The Earth's Water Cycle

Because water is the principal carrier for wastes, we need to consider how the "water cycle" works. The earth's water cycle is driven by energy from the sun. As we see in the sketch, water evaporates from streams, lakes, and oceans. It forms clouds and falls as rain or snow. Part of the water soaks into the ground, and part of it runs off, eventually reaching the rivers that carry it to the sea. Water also evaporates from the earth's surface and is released by vegetation. Thus water moves continuously from the atmosphere to the ground and back again.

A "water table" forms at some distance under the ground surface. It can be thought of as the top of the body of water that has settled into the ground. Its location varies from being near the surface close to the coast to deep below the surface in an arid inland region. Above the water table is the unsaturated zone, where the soil contains air and some moisture; below the water table is the saturated zone. The migration of wastes is much slower in the upper region. Water tables vary somewhat with time of day and considerably with the season. A well from which water is to be drawn must penetrate the earth at least to the water table, and preferably much deeper. Pumping water

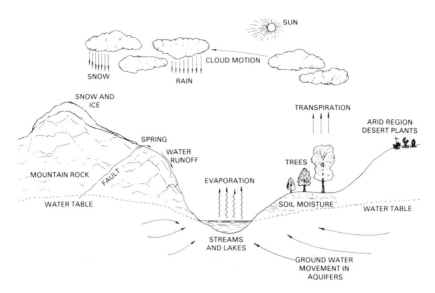

The water cycle. Precipitation falls on the earth, water flows within the ground, and evaporation completes the cycle.

from a well causes a tilt in the water surface, allowing the water to flow down the slope to match the extracted flow.

Water moves within the earth by means of aquifers, which are bodies of porous rock that readily conduct water. An example is a layer of gravel. As shown in the drawing, underground water generally tends to flow towards depressions just as surface water does but more slowly. If there is a great deal of rainfall and there are good aquifers, the water table will be near the surface. If there is little rainfall and the rock is impervious to water, the water table may be thousands of feet down. There may be more than one aquifer, separated by impervious material. Fortunately, aquifers tend to lie parallel to the earth's surface so that water flow is not directly upward.

Waste Retention in Geologic Media

The way dissolved materials are transported by underground water is very complex because of chemical and physical effects. Only by means of a computer program can the performance of a disposal facility be predicted accurately. Here is a simple example to illustrate how mathematical models are applied. Suppose that some tritium, the heaviest isotope of hydrogen, leaked from a waste container. Because tritium is practically the same chemically as ordinary hydrogen, water formed from it will move through the earth just like ordinary water. From the velocity of the water flow and the distance between the start and end of the motion, the time the tritium takes to get there can be easily calculated, as distance over velocity. That time would then be used to find out how much decay of the 12.3-year tritium has occurred during transit.

The geologic medium's ability to slow or trap radioactive material depends on the elements involved, on the chemical species, and on the electric charge of the ions in solution, as well as on the nature of the rock. A radioisotope such as cesium (30 yr) will move much more slowly than tritium. Its chemical form will cause molecules to become attached to soil particles or on pore walls in rocks as shown in the diagram. The molecules will later leave the surface and move along with the water, but the waste molecule motion has been delayed and the net speed reduced, an effect called "retardation." A retardation factor of, say, 35 means that the speed of cesium is only about 3 percent of the speed of water. The slow movement allows for decay that can reduce the radioactivity to a safe level by the time the cesium reaches public water supplies. Many of the chemical compounds formed from transuranic materials such as plutonium have very large retardation factors, and the final concentrations are low, even though the half-lives are long.

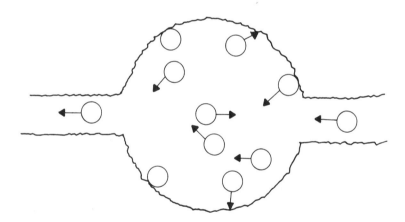

Retardation of waste chemicals in a geologic medium. The deposition of waste molecules on walls of pores in the rock delays migration.

Suppose that the waste were dissolved from the container at a constant rate over a definite period. As sketched, a "square pulse" of contamination would move along through the ground and pass a point that is farther along. The concentration of cesium, shown as the shaded area, is reduced by decay with time. Well water drawn from the aquifer at a distant point would be contaminated for a while, but would eventually clear up as the pulse moves on. A refinement of the computer model takes into account fluid-mixing effects and the nature of the geology that together cause some molecules to

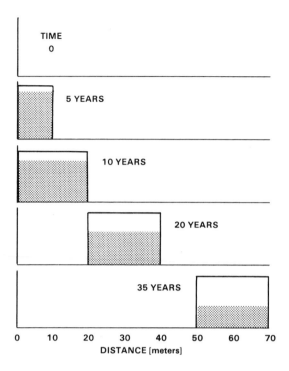

Movement of cesium-137 through the ground, showing reduction in concentration by radioactive decay. (Waste dissolving time 10 yr, waste speed 2 m/yr, half-life 30 yr.)

move more slowly while others move faster. This process, called "dispersion," reduces the peak radioactivity but lengthens the time the contamination is present.

The analysis recommended by the Nuclear Regulatory Commission is more complex than described here; it must take into account a variety of waste types and the many dissolved chemical species represented in the fission products and activation products from nuclear reactor operation. In addition, the ways the water is drawn from the ground and used by the public must be examined. Sophisticated and powerful computer programs are available to treat all of the processes. However, the programs must be tested for correctness, accuracy, and applicability to the actual situation.

Safety Objectives

A frequent question about any waste disposal site is, "How safe is it?" It would be good if one could reply, "Absolutely safe, forever," as if no radioactive material could ever escape. This is not seen as a realistic answer. Instead, as discussed in Chapter 6, recommendations of standards organizations are translated into dose limits by regulatory bodies. The effects of the manmade structures and the natural barriers that inhibit the transport of radioactivity are assessed as a *system*, the performance of which is compared with the dose limits. Such a performance assessment takes into account all available information and makes conservative assumptions where data are lacking.

In the design of disposal facilities a method called probabilistic risk assessment (PRA) is applied. It consists of identifying all of the possible events with their chances (and frequency) of occurring, and the consequences. The risk is calculated as the product of frequency and consequences.

It is important to decide the period of time over which calculations of radiation dose should be made. In the case of low-level waste, consideration of radioactive decay led to an NRC regulation of an institutional control period of 100 years. For protection of inadvertent intruders, those who enter the site after closure and may contact the waste, the period is 500 years.

In the case of high-level waste, which contains long-lived heavy elements such as plutonium and uranium, the period of isolation is much longer. A classic study by Cohen (see References) developed trends with time of the hazards of buried reprocessed spent fuel and buried intact spent fuel. Calculated were the contributions to dose of all of the fission products and the heavy element isotopes, and the probability of the waste escaping and being ingested by humans. The results are shown in the illustration. Also on the graph is the hazard of the original natural uranium from which the reactor fuel came. The curves cross at 14,000 years and 120,000 years, respectively. The paper notes that the number of fatalities over all time is small compared with those from coal burning.

Comparison of toxicities. The toxicity from intact spent fuel, reprocessed spent fuel, and the original ore from which the fuel came. (Adapted from B. L. Cohen, *Am. J. Phys.* 54(1), January 1986.)

The Environmental Protection Agency has specified that a high-level waste repository must meet established dose limits for a minimum of 10,000 years. That figure is based on both technical and policy considerations. Two choices for the compliance period were available: (a) the time to peak dose within the period of geologic stability, or (b) a fixed time period. EPA selected (b) with the 10,000-year figure, but EPA also expects DOE to report the expected peak dose, recognizing the uncertainty of such figures far into the future.

A great deal of attention has been given to the appropriate dose limit for disposal facilities. For reference, the allowed dose per year is 5 rem for a worker at a radioactive waste disposal facility. However, the ALARA principle discussed in Chapter 6 is to be applied to keep the actual doses lower than that figure. For any member of the public living near a low-level waste disposal facility, the annual dose above background according to NRC must be less than 25 mrem/yr for the whole body. However, in the case of high-level waste, as at Yucca Mountain, the annual figure set by EPA and adopted by NRC is 15 mrem/yr. This number serves as a performance specification. If we use 360 millirem as the average annual dose experienced by individuals in the United States, this limit corresponds to about 4 percent of the normal dose. Such an increase is regarded as insignificant in relation to the natural variations of dose by geographical location.

Natural Analogs

Models to predict waste migration can be partially checked against data collected on natural deposits of minerals. One of these is a uranium oxide deposit in Peña Blanca, Mexico. The ore is similar to the spent fuel in its corrosion characteristics. Another is a deposit of thorium in Brazil. Thorium behaves like an ion of plutonium. The rate of migration of thorium is found to be extremely small.

The most important analog is the "natural reactor" in Africa, at Oklo, Gabon. About two billion years ago, the concentration of ^{235}U in uranium was much higher than it is today. The half-life of ^{235}U is 7.04×10^8 years, while that of ^{238}U is 4.47×10^9 years. There was enough water present in the uranium ore bodies for a chain reaction to occur, with neutron bombardment, to produce fission products and plutonium. The process may have continued for hundreds or thousands of years at a low, self-limiting power level. Little migration has been noted of some radioactive species, especially ruthenium, technetium, and neodymium. Study of the amounts of different substances lost and remaining will help us to understand retention for the high-level waste repository planned for Yucca Mountain in Nevada. There are actually two types of natural reactors, one without organic materials present, the other with considerable bitumen (asphalt). The latter appears to bind wastes well, suggesting that bitumen be used in the repository.

It is an interesting exercise for the mathematically inclined reader to calculate what the atom percent of ^{235}U was two billion years ago, knowing the present 0.720 percent and the half-lives of the two uranium isotopes.

Long-Term Effects

Safe disposal of radioactive wastes involves selection of materials and geology ensuring that dose limits are met over a long period of time. These requirements are part of a broader issue of meeting energy needs in the near future and in the centuries thereafter. In the present era, energy stored in the form of coal and oil is abundant, but depletion of reserves in the future is considered inevitable.

The burning of fossil fuels in electric power plants and in vehicles for transportation adds large amounts of carbon dioxide and other gases to the atmosphere. These "greenhouse gases" create a blanket that traps heat radiation and contributes to general global warming. There is disagreement as to the magnitude of the temperature rise and its consequences, especially related to the possible rise in coastal water levels. At question is the relative effect of natural phenomena versus the effect of human activities. Indications are that wide fluctuation is normal, that the climate was more nearly tropical in the era of the dinosaurs, and that several ice ages have occurred in more recent times. One popular view, however, is that when in doubt, one should take every precaution to prevent harm.

Other effects of the release of chemicals by the consumption of fossil fuels are acid rain that pollutes bodies of water and kills vegetation, and depletion of the atmosphere's ozone layer, which protects us from skin damage by the sun's ultraviolet light.

There is uncertainty as to how to cope with all of these problems. Improvements in efficiency of energy use are possible but slow to implement, even in developed countries. There are needs for large amounts of additional energy in the developing world. It is not reasonable to expect conservation of energy if it implies reduction in its beneficial use. Greater use of solar energy and renewables is favored. However, the low energy intensity of sunlight and its variability limit its application, while the burning of firewood merely contributes to the atmospheric pollution.

In this context, nuclear power plants would appear to serve as a useful alternative to coal-fired plants. That idea is usually dismissed in discussions of the development-environment issue. With reference to the avoidance of global warming or the achievement of sustainable development, there is active resistance to conceding that nuclear power can be helpful. In the Kyoto Protocol and Earth Summit conferences (see References), the nuclear contribution to a clean environment was essentially ignored. Dismissed also was the possibility of using waste heat from nuclear plants to desalinate sea-water or clean up polluted water in developing countries.

Presumably, there are concerns about reactor safety, weapons pro-liferation, and the "waste problem." There is a tendency to reject nuclear without recognizing that there is a great deal of information available on performance of radioactive waste disposal systems, and that knowledgeable scientists and engineers agree that radioactive wastes can be safely managed.

CHAPTER 19

Disposal of Low-Level Wastes

The management of low-level radioactive wastes is evolving. Early disposal practices were inadequate; stricter regulations have been developed; and designs of facilities other than shallow land burial have been proposed. The problem of finding acceptable locations for disposal sites is generally believed to be more social than technical. This chapter reviews the history of low-level waste disposal and examines recent developments. Comments by Constance Kalbach Walker and John MacMillan on this chapter are appreciated.

Early Practices

Only after World War II did the problem of low-level wastes arise. New radioisotopes for medicine, research, and industry became abundant through irradiation in nuclear reactors. Their residues formed wastes, and they continue to do so. The second and much larger source was nuclear power generation, which produced contaminated materials and equipment.

The Atomic Energy Commission (AEC) maintained control of all nuclear activities for nearly a quarter of a century. The map shows the location of LLW generation, storage, and disposal sites. The AEC disposed of its wastes by near-surface burial. In 1963 the first commercial disposal site was established, and by 1971 there were six sites, located at West Valley, New York; Sheffield, Illinois; Maxey Flats, Kentucky; Richland, Washington; Beatty, Nevada; and Barnwell, South Carolina. The volumes and activities of LLW that have been placed as of May 31, 2001, in the six commercial sites are shown in the following table. The activity figures are for waste as emplaced, so the current values are much smaller. Note, the specific activity (Ci/ft^3) at Envirocare is extremely low. At all of these sites, the technique basically was to dig a trench, fill it with boxes and drums of waste, replace the excavated earth, apply some compaction, and form an earthen cap above the trench. As such, they were a cut above sanitary landfills, but below modern standards for disposal facility design.

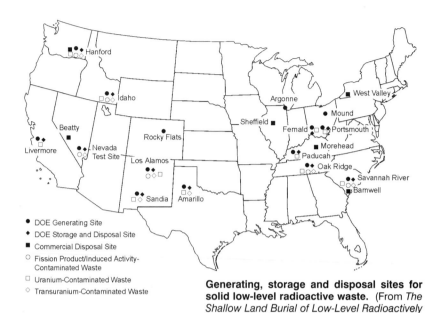

- ● DOE Generating Site
- ✦ DOE Storage and Disposal Site
- ■ Commercial Disposal Site
- ○ Fission Product/Induced Activity-Contaminated Waste
- □ Uranium-Contaminated Waste
- ◇ Transuranium-Contaminated Waste

Generating, storage and disposal sites for solid low-level radioactive waste. (From *The Shallow Land Burial of Low-Level Radioactively Contaminated Solid Wastes*. National Academy of Sciences, 1979.)

Volumes and Activities of Low-Level Wastes at Commercial Disposal Sites

(Source: Department of Energy, see References)

Site	Volume, millions of cubic feet	Activity, curies
Barnwell, South Carolina	27.07	10.34
Beatty, Nevada	4.85	0.64
Envirocare, Utah	6.26	0.0023
Richland, Washington	13.53	4.01
Maxey Flats, Kentucky	4.78	2.40
Sheffield, Illinois	3.12	0.062
West Valley, New York	2.72	1.27
Total	62.33	18.72

Shallow Land Burial

The disposal of LLW in a trench slightly below the surface is called shallow land burial, in contrast to deeper mined-cavity disposal as practiced for high-level wastes. The diagram shows the conventional arrangement, with the wastes placed a suitable distance above an aquifer and the site fenced to prevent entrance. There has been mixed success with this disposal technique. Three of the commercial sites (West Valley, Sheffield, and Maxey Flats) developed leaks and were closed, while the remaining three have operated satisfactorily.

Shallow land burial of low-level wastes. (Courtesy of R. F. Weston, Inc.)

Three principal types of failure have been noted.* The first type is simple erosion by surface water, which exposes waste containers to the elements. The second type occurs when the wastes are loosely packed and will compress under the weight of dirt. Then the cap subsides, pockets of water appear, and water percolates into the waste, eventually leaching out the radioactive material. The third type is the "bathtub effect," in which a leaking cap lets water into an excavated cavity with nearly impermeable walls. The wastes are immersed in water for long periods, the containers corrode, and the wastes dissolve. The "bathtub" fills and overflows, carrying contaminated water to the environment.

Accounts differ widely as to the amounts of radioactivity released and the resultant hazard to the public. It is a fact, for example, that tritium and strontium-90 have been measured at Maxey Flats. One estimate is that the resulting dose was less than 1 percent of that from natural background. Statements by others suggest that the hazard was much greater.

The design of such earlier facilities was clearly inadequate. The problems were due partly to insufficient investigation of geologic features before sites were selected. Contributing factors were the loose packing of wastes, the presence of liquids in the waste as received for disposal, and the poor design of caps to exclude water. Operating companies claim that they followed the federal regulations existing at the time. Hindsight reveals the primitive nature of regulations on site selection, waste packaging, and facility design.

The remaining sites (Richland, Beatty, and Barnwell) had problems of a different type. Poor packaging of wastes by shippers resulted in contamination of the disposal area when the packages arrived at the sites. At one location, contaminated equipment was stolen. The three states were concerned with the injustice of having to receive the bulk

---

*See *Proceedings of the Symposium on Low-Level Waste Disposal*, Vols. 1-3, NUREG/CP-0028, 1982-3.

of the wastes from the entire nation. As a result, the governors of the states threatened to close the facilities, leaving no place for the LLW to go. This prompted Congress to pass in 1980 the Low-Level Radioactive Waste Policy Act, which placed responsibility for waste disposal on the states producing the waste. The act also recommended that groups of states establish regional facilities. Several interstate compacts have been arranged, as discussed in Chapter 22.

Site Selection

Experience at the closed disposal sites led the Nuclear Regulatory Commission to develop a new set of siting rules. The result after several years of work was the regulation titled "Licensing Requirements for Land Disposal of Radioactive Waste," Title 10 of the *Code of Federal Regulations*, Part 61, commonly referred to as 10 CFR 61. The regulation requires low-level wastes to be classified by the generator into one of three categories—A, B, or C, taking account of half-lives and concentrations of the nuclides in the wastes. Generally, Class A wastes are of lower activity than Class B or Class C. The latter require special attention, as will be discussed in Chapter 22. Regulation 10 CFR 61 stresses the need for careful assessment of geology, hydrology, and other features. Some key requirements are

(a) the site geology should be simple enough to admit mathematical modeling and computation,

(b) location of the site should be far from housing or commercial development that would affect underground water flow,

(c) there should be no significant underground resources whose extraction would affect site performance,

(d) the sites should be well-drained to reduce infiltration of water,

(e) there should be little chance for water flow across the site that would erode the surface,

(f) the water table at the site should be deep enough to prevent immersion of wastes,

(g) there should be no springs in the site that could bring contamination to the surface,

(h) the area should not be subject to volcanic action, earthquakes, landslides, or excessive weathering,

(i) nearby activities should not affect performance or monitoring.

The process by which a site is selected consists of three parts. The first is a general survey of the region or state, examining existing records on the geology, hydrology, meteorology, seismology, population distribution, land use, environmental features, and cultural aspects. The second is an interaction between the siting agency or company and the communities that may wish to serve as host to the facility or at least will find it acceptable with appropriate compensation. The third is a series of measurements, including a limited number of test borings, leading to a site characterization, which is a catalog of features that relate to the siting regulations.

The site characterization study following NRC guidelines must be carried out before a license to operate can be issued. Such an investigation must include data taken over a full year. Once a license is issued, the disposal site can be opened to receive wastes and will operate for about 30 years, during which regulatory surveillance is provided by the NRC or the cognizant state agency. At the end of the operating period, the site may be closed, meaning that no additional wastes are received. However, institutional control must continue for at least an additional 100 years. This involves measurements of waste migration and remedial action as needed. After institutional control is released, the site may be used safely for any purpose. The design is expected to protect the public for at least 500 years after closure. The performance specification of no more than 25 mrem/yr to any member of the public must be continued.

Greater Confinement Disposal Alternatives

The NRC has suggested several techniques by which the 10 CFR 61 requirements might be achieved, starting with conventional shallow land burial with sloped trench walls. Use of a narrow trench with vertical walls cuts down on the radiation exposure to operators as they place the wastes. Containers can be stacked in a regular pattern to reduce the amount of potential void space. Cubes or hexagons leave essentially zero voids. Square 55-gallon drums are now available commercially. Layered waste disposal puts material of high activity in the bottom of the trench to protect the inadvertent intruder. The NRC recommends several moisture barriers as part of the cap system. It also suggests a design for engineered barriers against intrusion.

The three disposal facilities at Richland, Beatty, and Barnwell did not experience difficulty with waste migration and were regarded as successful even though they are examples of shallow land burial. Those in the West are in arid regions. Barnwell is in a humid region with heavy rainfall but is located where much of the soil is clay, which prevents water intrusion and filters contaminants effectively.

Despite these favorable conditions, publicity about the now-closed LLW disposal sites has given shallow land burial a bad reputation. The public is concerned about reactors, radioactivity, and radiation in general, and shallow land burial in particular. Consequently, some states and compacts have prohibited its use or have required an "alternative waste disposal technology" that involves "greater confinement disposal." Several design concepts that incorporate "engineered barriers" have been proposed. Generally, these provide additional protection against waste migration.

The Nuclear Regulatory Commission finds that a facility designed to meet its performance specifications is satisfactory without any added features. However, the public often seems to have one or more of these different views: (a) the limiting dose should be nearer zero or even should be actually zero, (b) some unexpected event might change the system from the one analyzed, (c) the knowledge of underground water flow is inadequate, and (d) there may be human error in the analysis, design, construction, and operation of the facility. It is hard to convince people otherwise, and in some states and interstate compact regions, legislation on additional protection has been passed to make a waste disposal facility acceptable.

In this section, some advantages and disadvantages of several of the proposed disposal concepts are noted. Referred to are the accompanying sketches, kindly provided by EG&G Idaho, a former operator of the Idaho National Engineering and Environmental Laboratory.

Intermediate-depth disposal is similar to shallow land disposal except for the greater trench depth and for the thickness of the cover, which may be more than 30 feet. The cover minimizes water infiltration, provides ample shielding against radiation, and discourages future intrusion.

Intermediate-depth disposal. Similar to shallow land burial except for deeper trench and thicker cover. (Courtesy of Idaho National Engineering and Environmental Laboratory; artist David Combs.)

Mined-cavity disposal consists of a vertical shaft going deep in the ground, with radiating corridors at the bottom. This is quite similar to the planned disposal system for spent fuel and high-level wastes from reprocessing. This method is not likely to be used in the United States because of the greater expense. Note, however, Finland uses cavities cut in bedrock for LLW disposal. Deep disposal can be used for wastes not suitable for near-surface disposal.

The *belowground vault* provides a concrete barrier to infiltration of water and to the escape of radioactive materials. Reinforced concrete forms the walls, floor, and roof, with waste containers stacked closely together and the spaces between filled with sand or gravel. The vault enhances stability under the pressures of overburden and protects the waste against the effects of earthquakes. It sits on a porous pad to allow any water that enters to drain away. The structure also provides gamma-ray shielding for those at the surface and minimizes future intrusion. Concrete with an average life of around 500 years is required, using special mixes and methods of preparation. The earthen cap has several layers: (a) vegetation that is hardy but has shallow roots, (b) topsoil, (c) rocks to prevent root entrance, (d) sand and gravel to carry water sideways, (e) clay to provide a barrier nearly impervious to downward water flow, and (f) sand and gravel to divert water. The cap also provides shielding and deters intrusion.

Belowground vault. Concrete structure with earthen cover. (Courtesy of Idaho National Engineering and Environmental Laboratory; artist David Combs.)

Modular concrete canister disposal consists of placing individual waste containers within concrete cylinders called "canisters," with grout filling the spaces between containers. The canisters are then buried below grade in a shallow land site, with sand between them for stability and with the usual earthen cover.

Earth-mounded concrete bunkers, used by the French at a recently filled facility, have the favorable features of the belowground vault and the modular concrete canister. Well below the natural grade a concrete bunker is constructed to hold the more radioactive Class B and C wastes. The mound above, called a "tumulus," is formed by a stack of Class A concrete containers, of lower activity. A rounded earthen cover allows runoff, prevents infiltration, and provides radiation shielding.

Modular concrete canister disposal. Waste placed in canisters that are placed in trenches. (Courtesy of Idaho National Engineering and Environmental Laboratory; artist David Combs.)

Earth-mounded concrete bunker. Class B and C wastes in bunker with Class A wastes in mound above. (Courtesy of Idaho National Engineering and Environmental Laboratory; artist David Combs.)

Shaft disposal uses concrete for walls and a cap. A series of individual holes are drilled in the ground and concrete cylinders are installed, into which waste containers are placed. The method is seen to be a variant on either the modular concrete canister or the below-ground vault.

The *aboveground vault* is a simple reinforced concrete box into which containers of waste are placed, with sand between containers to ensure stability. A concrete slab is poured in place above each cell after it is filled with waste containers. If necessary, the waste containers could be retrieved. Thus the concept is more nearly "storage" than "disposal." These words have significant overtones in terms of public acceptance. Storage is favored by those concerned about the presumed hazard of the burial of waste with future concealed problems; disposal is favored by those who believe that later generations should not be burdened with long-term custodial responsibility.

Aboveground vault. Concrete building without earthen cover. (Courtesy of Idaho National Engineering and Environmental Laboratory; artist David Combs.)

As no earthen cover is provided, the concrete roof is exposed to the effects of weather, including erosion, freeze-thaw cycles, and acid rain. Of course, deterioration of the structure could be noted, and repairs made, but surveillance would need to be extended, perhaps indefinitely. Analyses show that if the aboveground vault is unattended, the concrete would deteriorate and the wastes eventually would be dissolved and transferred by surface water, giving an unacceptable radiation dose. The concept's features are not particularly compatible with the requirements on protection of an inadvertent intruder, who encounters the facility beyond the end of the 100-year institutional control period. For these reasons, the NRC does not favor this method of disposal, as noted in Volume 3 of the series of NRC reports on alternatives (see References).

A concept that combines features of others and goes considerably further in terms of protection is the *integrated vault technology*, designed by Chem-Nuclear Systems, Inc., intended for use at several U.S. disposal sites but never implemented.

The design borrows from the technology used by the French at Centre de l'Aube. As sketched, earth is mounded above each module, with a sloping multi-layer cap to help runoff. A plastic layer is added to reduce infiltration of water. The facility uses reinforced concrete vaults, into which concrete canisters called "overpacks" are placed, each holding one or more low-level waste containers. Overpacks are of two types. One is cylindrical, into which a number of 55-gallon drums may be fit, with spaces between filled with grout. Alternatively, a single high-integrity container can be inserted. The other is rectangular to accommodate square liners or a set of boxes. The use of overpacks in addition to the module vault provides structural stability and an additional layer of protection. All classes of waste are put in overpacks. The higher activity B and C wastes are placed in the bottom of the vault, with A wastes above and beside them. This arrangement provides shielding and protection of future inadvertent intruders. In contrast with other concepts, spaces between overpacks are not filled with sand or gravel but are left open

Waste containers are placed in concrete overpacks and sealed with grout.

Drums or Cylindrical Shipping Containers

Square or Rectangular Shipping Containers

Overpacks are placed in reinforced concrete modules which are closed with a reinforced concrete roof.

Integrated vault technology. (Courtesy of Chem-Nuclear Systems, Inc.)

to allow free drainage of any infiltration. The vaults sit on porous pads for drainage, with the bottom of the disposal facility well above the water table. Each vault has a drain to a monitoring system, located in a tunnel between adjacent rows of vaults. The tunnel allows access for inspection, sampling for possible water drainage from the vaults, and measurement of any radioactivity. Such a facility is obviously much more expensive than simple shallow land burial or other variants but would meet stringent requirements, even for a humid region.

To visualize how a facility that is fully underground provides protection for the public start with the sloped multi-layer cap that allows runoff of a large fraction of rainfall. Most of what infiltrates is drawn sideways by the sand-gravel layer, and the combination of plastic and clay beneath prevents infiltration. The concrete roof and its contents thus remain dry for many years. During the 100-year institutional control period, which can be longer if desired, remedial action can be taken if significant amounts of water enter any vault or if radioactivity is detected by the monitoring system. The plastic in the cap is assumed to deteriorate after a number of years, and some water will enter, contacting the vaults. Eventually cracks will appear in the concrete roof and walls. Moisture can then contact the outside of the concrete overpacks, but because of the open spaces between them, they are not exposed to standing water. Over a period of many hundreds of years, the overpacks are assumed to deteriorate, gradually allowing water to contact the waste containers. The concrete roof and overpacks still effectively warn any inadvertent intruder and provide a physical barrier to exploration. By this time, much of the radioactivity has disappeared by decay. Finally, water diffuses into the waste form, slowly dissolves some of the chemicals of the waste, and diffuses out into the vault, to be carried away in the ground beneath the facility. As in any of the disposal concepts, each chemical migrates with a speed slower than that of the underground water by an amount called the "retardation factor," values of which vary from one to thousands. Knowledge of the water flow patterns and velocity allow prediction of the amount of decay in transit through the buffer zone, which is a minimum distance (e.g., 1000 to 2000 ft) between the facility and the nearest public water well. The facility design and the geologic environment act as a *system* that stores and isolates the wastes and delays the release of radioactivity. Concentrations of activity are not zero but are expected to be well below the regulatory limit during operation and for all of time after the facility is closed.

The cost of disposal of low-level waste continues to increase for at least two reasons. First, legislators and regulators have responded to public demand by requiring more secure and complex disposal facilities. Second, in the interest of economy, generators have actively sought to reduce volumes of waste to be shipped and disposed. As an example of disposal costs, consider those at the Barnwell facility, regulated by the South Carolina Energy Office (see References). Base charges range from $4.40/lb to $7.70/lb, depending on the volume of the shipment. An extra charge of $1/lb is assessed if the waste is of biological origin. Further charges depend on the number of milli-curies and the dose rate. A portion of the income from waste disposal goes to the community or state that serves as host; another part goes into a fund for remediation if needed; and finally money is collected for the design and development of the next facility to serve the state or region. Generators pass the costs on to the customers who use nuclear electricity and to the beneficiaries of medical and research applications. The public ultimately pays the bill, of course.

Special Classes of Wastes

Most low-level wastes come from nuclear reactor operations and radioisotope applications, regulated by the NRC. One type that is the responsibility of states is NARM (naturally occurring and accelerator-produced radioactive material). The EPA rule 40 CFR 193 will address such substances as well. Part are accelerator targets and other equipment. Others are soils containing natural uranium or radium. The latter element was produced naturally in large quanti-ties in earlier times, as described in the *Scientific American* article "The First Nuclear Industry" (see References). Radium contamina-tion exists at former plant sites. Such wastes may be disposed of as LLW or sent to a desert disposal facility in Utah operated by the com-pany Envirocare.

Mixed wastes (also called "mixed low-level wastes") are those con-taining both hazardous chemicals and radioactive substances. Haz-ardous wastes are defined as materials that are toxic, corrosive, inflammable, or explosive. They contain specific elements such as lead and mercury, pesticides such as DDT, and cancer-producing compounds such as PCBs and dioxin.

The disposal of hazardous wastes is regulated by the Environmen-tal Protection Agency under the Resource Conservation and Recovery Act (RCRA) while radioactive wastes are controlled by the Nuclear Regulatory Commission under the Atomic Energy Act. Thus it has been necessary to establish consistent dual rules by agreement between the agencies. The EPA and NRC provide a formal procedure by which one can decide whether a certain material is mixed waste. They also give written suggestions for the design of a disposal facil-ity that meets both agencies' requirements. Included are double liners and leachate recovery equipment for EPA and waste isolation and intruder barriers for NRC. It is estimated that only a few percent of commercial and institutional low-level wastes are in the category of mixed wastes. Their disposal, however, is estimated to be very expensive, as high as $15,000 per cubic foot. Because the total volume of mixed wastes is relatively small, they do not pose a major problem, but their management has to be addressed.

Generators of mixed waste have a sequence of possible options: (a) to select alternative materials and processes that do not involve hazardous organics or metals, leaving only radioactive materials to deal with, (b) to use operational control to reduce the radioactive content below NRC concerns, leaving only the hazardous component to deal with, (c) to store wastes for decay if half-lives are short enough, (d) to ship wastes to a processing facility which can treat them to eliminate the hazardous chemical property, leaving only the radioactive component, and (e) to ship the wastes to a jointly regulated facility.

States and compacts in turn have options. One is to build and operate a separate special disposal unit for mixed wastes, one that meets all regulators' requirements. Another is to seek relief by requesting the Department of Energy to accept mixed wastes for treatment and disposal; commercial wastes are a relatively small fraction compared to DOE's accumulation. A National Mixed Waste Profile was developed in 1990, consisting of an inventory of amounts produced and in storage, and an identification of facilities capable of processing them. A description of the joint effort of NRC and EPA is found on a Web site (see References). Such data are useful for defense waste disposal, to be discussed in the next chapter.

"Greater-than-Class-C" (GTCC) wastes are in the category of low-level wastes but differ in two ways. They either have a higher activity than the upper limit of the NRC's Class C or contain larger amounts of transuranic materials than allowed. Examples are activated metal wastes from reactor decommissioning and transuranic wastes from nuclear fuel testing or uranium-plutonium mixed oxide fuel fabrication. Such materials cannot be given near-surface disposal. The Department of Energy is committed to accept GTCC wastes for storage and disposal along with high-level wastes. The volume of such wastes generated through the year 2020 is estimated to be relatively small.

CHAPTER 20

Disposal of Defense Wastes

Nuclear technology seems relatively new, but the year 2002 was the sixtieth anniversary of the startup of the first nuclear reactor. For the latter half of the twentieth century, radioactive defense wastes have been generated, first in the World War II atom bomb project, and ever since in producing materials for nuclear weapons and in reprocessing fuel from nuclear-powered naval vessels. Given the national policy to maintain a strong nuclear defense as part of the Cold War, the production of wastes was inevitable.

The defense products have always had the highest priority, and wastes have been handled in ways that seemed appropriate at the time. Thus decisions were made to store waste in temporary single-walled underground tanks, as at Hanford, or to bury them in ways regarded as inadequate by modern environmental standards, as at Idaho Falls. Waste management plans must include remedial action as well as current needs.

The complexity of the defense waste problem can be seen in the mere listing of its many dimensions. First are the sources: from treatment of fuel from several types of reactors—those for production of plutonium and tritium; naval reactors; and test reactors—and from weapons production and research programs. Second are the geographic locations of the defense sites that produce waste; the major ones are Hanford, Savannah River, Idaho Falls, Oak Ridge, and Los Alamos. Third are the types of waste—high-level (HLW), low-level (LLW), and transuranic (TRU). Fourth are the forms in which the wastes are currently stored. Fifth are the various forms into which the waste can be put—capsules, cement-waste mixtures, and glass-waste mixtures. Sixth is the status—buried, stored, or freshly produced. Seventh are the possible ways to dispose of the waste—stored for future decision, immobilized in its present location, shipped to a special repository, or prepared for disposal in the planned commercial repository. The stated purpose of the defense waste program of the Department of Energy is to provide safe and economical management of waste to protect the environment and ensure safety and health of workers and the public. In the following sections, the amounts of wastes to be handled, some of the key technical processes, and the strategy under way are discussed.

Amounts of Wastes

An appreciation of the magnitude of the task facing the Department of Energy can be gained from inspection of tables of data that give volumes of the main types of waste at various locations.

For high-level wastes, estimated amounts as of 1999 are shown in the next table, extracted from a DOE report of April 2001 (see References). Included is the number of canisters of vitrified waste, which is waste encased in glass. Projections on inventories far into the future are found at the DOE Environmental Management Web site (see References).

High-Level Wastes at DOE Facilities

Location	Volume, m^3	No. of canisters
Hanford	199,902	—
Savannah River	130,051	719
Idaho Falls	9357	—
West Valley	109	241
Total	339,419	960

Volumes of transuranic wastes as of 2000 are shown in the table below. In the United States, TRU is found at 28 locations.

Transuranic Wastes at DOE Facilities

Location	CH, m^3	RH, m^3
Rocky Flats	5307	—
Lawrence Livermore	295	—
Idaho Falls	64,878	85
Los Alamos	9213	99
Nevada Test Site	665	—
West Valley	73	467
Mound Lab	247	—
Savannah River	10,991	—
Oak Ridge	963	1342
Hanford	16,100	200
Other sites	195	54
Total	108,927	2246

CH = contact handling
RH = remote handling

In addition, there is a large volume of buried TRU, mainly at Los Alamos, Hanford, and Idaho Falls. The total is estimated to be 497,000 m^3. Most of this waste will probably be left alone but monitored.

The principal amounts of low-level waste as of 1999 are shown in this table.

Low-Level Wastes at DOE Facilities

Location	Volume, m^3
Lawrence Livermore	1548
Rocky Flats	12,208
Idaho Falls	2188
Paducah	6373
Sandia	599
Brookhaven	611
West Valley	16,279
Portsmouth	17,433
Savannah River	15,276
Oak Ridge	47,371
Other sites	960
Total	120,846

The main volumes of mixed low-level waste (MLLW) are shown in the following table.

Mixed Low-Level Wastes at DOE Facilities

Location	Volume, m^3
Lawrence Livermore	433
Rocky Flats	12,242
Idaho Falls	4058
Paducah	4176
Los Alamos	325
Portsmouth	9059
Savannah River	2258
Oak Ridge	24,271
Hanford	9171
Other sites	487
Total	66,480

Data on the amounts of mill tailings were given in Chapter 15. Not included in the above tabulations is a category called "contaminated environmental media," consisting of water, soil, sediment, sludge, and debris.

The total volume of the wastes listed in the tables is 637,918 m^3. It is difficult to visualize such a size, just as it is for a million barrels of oil or a billion dollars. However, all of the wastes could be housed in a cubical building occupying a typical city block. Eventually all of this material must be processed and secured.

Waste Form

The wastes from reprocessing to remove plutonium and uranium were first generated in 1944. They are composed of a great variety of chemicals, including fission products; metals such as iron, manganese, and aluminum; sodium salts; organic materials; and special radionuclides strontium-90 and cesium-137. They are in the form of sludges (of consistency from thin paste to peanut butter), solid salt cake (from jelly to concrete), and liquid, all traditionally stored in underground tanks, especially at Hanford and Savannah River.

As seen above, there are enormous volumes waiting to be processed into a safer form. The diagram shows a waste tank with its various openings and measuring devices. The 149 early tanks at Hanford were single-walled, and in the 1960s, about 67 of them developed leaks. As much as a million gallons of contaminated liquid escaped. The nature of the soil in southeastern Washington State has provided protection to the public. A National Academy of

Defense waste storage tank. (From *The Safety of Nuclear Reactors and Related Facilities*, WASH-1250, Atomic Energy Commission, July 1973.)

Sciences report (see References) notes the sorptive character of the layers of gravel, sand, silt, and clay, and states that the radioactivity has been essentially immobilized. Under the report's worst-case scenario, radiation doses at the Columbia River are predicted to be very small. Despite that optimistic opinion, the Hanford Site, through DOE's Office of River Protection, has maintained vigilant monitoring of the areas near the Columbia River. These studies determine the need for remediation of groundwater and the vadose zone (that between the groundwater and the surface) to minimize contamination of the river by radioisotopes such as cesium-137 and strontium-90. Later 28 double-walled tanks were built. On detecting a leak in the inner wall, there is ample time to transfer the contents to an intact tank. An extensive ground-monitoring system tracks movement of contamination from past leaks. This includes sampling of the contents of the unsaturated zone as well as water measurements below the water table.

Considerable attention has been given to the possibility of an explosion in the Hanford waste tanks. To extract ^{137}Cs and ^{90}Sr, the chemical ferrocyanide was used. Its residue can react with other chemicals to generate heat. Also, hydrogen is released in certain reactions and can build up to hazardous concentrations. To forestall such an event, a large 20,000-lb test pump was developed. It stirs the mixture in a tank, allowing hydrogen to come out continuously rather than accumulating. Tests on the pump in Tank 101-SY have been regarded as very successful. A permanent pump is planned, with the test pump moving to other tanks.

Tank storage has to be regarded as temporary, and it is necessary to process the wastes in preparation for ultimate disposal. This section describes the steps required at Hanford and Savannah River. First is characterization, which is the sampling and analysis of the materials to determine their physical, chemical, and radiological nature. In the early days, records on radioactivity were better than those on chemical content. With 177 tanks to deal with at Hanford, extracting and analyzing even a few samples per tank is a major task. The second step is retrieval. Pumping will work well for liquid, but special excavating tools are required for hardened material. One device called the "Tank Crawler" is similar to a robot. It is inserted into a tank to dislodge solids, using remote control. Third is pretreatment, in which high-level waste is separated from low-level waste and from nonradioactive materials.

The high-level component of the waste is to be immobilized by mixing it with material that forms a solid resistant to attack by water or the solutions that may be present underground. The substance should contain an adequate fraction of waste as impurity and still remain strong and uniform in composition. It should not be damaged by heat or radiation from fission product decay.

Glass has been studied extensively and is the material selected for most defense wastes. Glass has several favorable features: it mixes well with wastes of various compositions, casts easily into proper form, conducts heat well, and resists attack at low temperatures. Some other properties are its noncrystalline (amorphous) form, in contrast with substances such as ice, salt, and sugar. Use of the word "crystal" to describe a form of glass is a misnomer. It is in a vitreous condition as a supercooled liquid that is subject to fracture. Under excessive heat, stress, or radiation, glass could devitrify and break into pieces. The smaller the fragments, the more easily the glass is leached (dissolved) by water. The principal chemical in glass is silicon, an ingredient of sand. Typical commercial glass as used in bottles or windowpanes has this composition:

Compound	Percent
Silica (silicon dioxide, SiO_2)	71.5
Soda (sodium oxide, Na_2O)	14.0
Lime (calcium oxide, CaO)	13.0
Alumina (aluminum oxide, Al_2O_3)	1.5

The glass-waste mixture used by the French has a composition that is more like Pyrex®, the heat-resistant glassware used for cooking made by Corning. It is a borosilicate type, in which boron oxide replaces lime. When fission products are included, the glass is black rather than transparent and has this composition:

Compound	Percent
Silica	42.5
Soda	14.0
Boron oxide	17.5
Fission product oxides	13.0
Alumina	8.5
Other	4.5

Much research has been done on glass as a waste form. The influences of mechanical stress, radiation, and corrosion have been examined. Of special interest is the effect of water, saltwater brine, and bitterns, which are brackish residues after salt has crystallized out of saltwater. The amount of leaching depends on the element. For example, the amount removed per day in micrograms per square centimeter is 1 for cesium, 0.1 for strontium, 0.01 for cerium, and 0.001 for ruthenium.

The vitrification process that has been developed for HLW by the Pacific Northwest National Laboratory starts with liquid waste. It is mixed with glass-forming chemicals, and the combination is introduced into a melter, as shown in the drawing. The mixture dries and the resultant solid melts. The equipment is operated remotely, behind a concrete wall 4 feet thick. At the temperature of 1150°C, organics are destroyed and inorganics dissolve in the glass. Heat is generated in the molten glass by a large electric current. A turntable allows a canister to be filled with glass, then moved around to cool, bringing up an empty canister. Gas from the process is treated to remove water, heat, and radioactivity such as carbon-14, ruthenium, and iodine-129. The canisters are welded shut, decontaminated, and stored in racks or air-cooled vaults, awaiting disposal in a repository.

At Hanford, a major cleanup effort consists of removal of wastes from tanks and immobilizing the wastes in glass. The company CH2M HILL Hanford Group has the task of maintaining the existing tanks, transferring material to safer storage, and performing pre-treatment of wastes. Bechtel National started construction in 2001 of the large Waste Treatment Plant where vitrification will take place, starting in 2007. The total cost is to be over $4 billion and will employ 4300 people. The facility's melters will use most of the 62.5 megawatts electrical supply. Two streams—high-level waste and low-activity waste—will be vitrified separately. "Logs" of low-activity glassified waste will be placed in lined underground trenches while high-level glassified waste will be stored in existing vaults, awaiting shipment to a repository. Progress reports on the project are available on the Web (see References).

Melter and turntable designed for vitrification plant at Hanford. The molten waste-glass mixture solidifies within the metal canister. (Courtesy of the Pacific Northwest National Laboratory.)

In 2002 the Secretary of Energy announced a draft of *Performance Management Plan for the Accelerated Cleanup of the Hanford Site*. Included is a cleanup of the bulk of the site near the Columbia River, completing the tank program by 2035 rather than much later, storing plutonium as well as capsules of cesium and strontium, and using an old chemical separation building for waste disposal (see References).

Vitrification Plant

At the Savannah River Site 34 million gallons of high-level waste with an activity of 538 megacuries are being treated. At this site, Westinghouse Savannah River Company has operated the world's largest production version of the vitrification process since 1996. This process is located in the Defense Waste Processing Facility (DWPF). The facility cost was $4 billion. By May 2002 it had generated more than five million pounds of glassified waste. The facility in which the vitrification is done is 117 feet wide, 94 feet high, and 360 feet long, designed to withstand earthquakes and tornadoes. It features remote control, closed-circuit television, and a monitoring computer system. The process treats two kinds of waste—precipitate and sludge, as shown in the flowsheet.

The Defense Waste Processing Facility turns radioactive precipitate and sludge into a glassy solid using the vitrification process.

Borosilicate glass frit is added to waste; the slurry is evaporated, and sent to the melter operating at 2100°F. The resulting liquid is poured into stainless steel cylinders 10 feet long and 2 feet in diameter. Of the weight of 5000 pounds, only 60 to 90 pounds is actual waste. Canisters are decontaminated, sealed, and stored, awaiting disposal at a HLW repository. At the rate of production, it is expected to take about 20 years to complete the task of processing the contents of 49 underground tanks of HLW. Additional details on the process are found in the References.

The low-level salt solution is mixed with cement and fly ash (as from burning coal in power plants) to form a grout. This is poured into large concrete vaults hundreds of feet long to solidify as "salt-stone."

In Situ Vitrification

A novel spinoff of the above solidification process that may be useful in stabilizing certain types of radioactive waste is "in situ" (in place) vitrification (ISV). Developed at the Pacific Northwest National Laboratory, ISV converts contaminated soil or other earthen materials into a glass-like solid by electric joule heating.

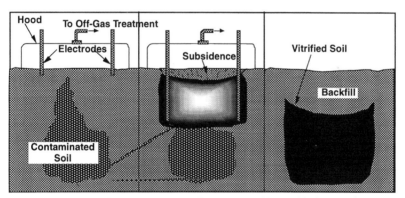

In situ vitrification. (Courtesy of the Pacific Northwest National Laboratory.)

Four electrodes in a square array are placed slightly into the soil to be treated. A starter path of graphite plus glass is laid down between the electrodes. Four thousand volts of electric power is applied to the starter path; this heats up and melts the adjacent soil, which, upon becoming molten, becomes electrically conductive. The molten soil then becomes the conductor for the heating process. Up to 3.5 megawatts of power (roughly the amount consumed by a good-sized hotel) are applied, and the melt is heated into the 1600 to 2000°C range. The molten mass grows downward and outward past the electrodes until the desired volume is encompassed. The electrodes are progressively lowered while the melt grows downward.

A glassy and microcrystalline solid similar to volcanic obsidian results upon several months of cooling. Individual melts as large as 1400 tons and 22 feet deep have been made in a single setting of electrodes. A significant volume reduction of 25 to 45 percent occurs with most soils because of the elimination of void volume. The vitrified product is totally free of organic contamination because organics are thermally decomposed by exposure to the melt temperature. Heavy metals like plutonium, uranium, and lead become permanently immobilized by chemical incorporation into the vitrified material. Leach rates are significantly smaller than those for vitrified HLW, primarily because of the higher silica content associated with contaminated soil applications. It is estimated that the vitrified product will retain its burden for geologic time spans.

In addition, ISV has been developed for soils contaminated with hazardous chemicals. The process has proven effective for a broad range of organic and heavy metal contaminants. It is particularly competitive for more difficult applications involving mixtures of these waste types, or where application in place is important. Sublicensed to Geosafe Corporation, the technology is being marketed worldwide, and commercial large-scale applications began in mid-1993. Current projects deal with mercury, pesticides, dioxin, pentachlorophenol (PCP), polychlorinated biphenyl (PCB), plutonium, uranium, beryllium, and other organic and heavy metal wastes.

Disposal Strategy

Many types of defense wastes at many sites need to be disposed of in accord with requirements of the Environmental Protection Agency. Priorities must be established to accomplish the task at reasonable cost. For a number of years, the Department of Energy has been implementing a long-range management plan. In 1989, the program was broadened in scope to include environmental restoration, and subsequently several five-year plans were issued. An important feature is prioritization, in which problem sites are attacked in order of importance and urgency. The high-level and transuranic wastes at three sites will be disposed of in sequential order. The Savannah River Plant has the greatest fraction (59 percent) of the DOE radioactivity. With its wet climate and small depth to the water table, Savannah River has the greatest potential for release of buried wastes. Those will be handled first. Information so gained will be useful for the Hanford Site, and in turn for the Idaho National Engineering and Environmental Laboratory.

Transuranic waste poses a special problem because of its peculiar history. Before 1970 it was buried as LLW. In 1970 TRU was defined as wastes containing 10 nanocuries per gram of alpha-emitters (redefined in 1982 as 100 nCi/g). Since 1970 it has been kept separate from LLW and stored in underground bins. Three categories require different treatments: contact-handled (CH-TRU) with surface dose rate less than 200 mrem/hr; remote-handled (RH-TRU); and special (SP-TRU), which includes large contaminated pieces of equipment. About 99 percent is the CH type.

The general plan is to send TRU to the Waste Isolation Pilot Plant, described below, once it has been checked for certain properties. The TRU is transported in the special container called TRUPACT II, as described in Chapter 17.

The low-level waste from the three defense sites and other DOE sites will be buried in some form of greater-confinement disposal. Choices include those for commercial low-level waste (Chapter 19) along with lined trenches, caissons, or augured holes.

Waste Isolation Pilot Plant

In 1957 a committee of the National Academy of Sciences and the National Research Council reported the results of their study entitled *The Disposal of Radioactive Waste on Land*. The group said that it was "convinced that radioactive waste could be disposed of safely in a variety of ways and at a large number of sites in the United States." They indicated that waste storage in tanks was safe and economical for the present but that their first choice was solidification followed by underground burial. Their second choice of permanent disposal was use of silicate bricks in surface repositories or dry mines. Their main recommendation was that "disposal in salt is the most promising method for the near future." The salt recommended was natural sodium chloride in rock form, deposited long ago when oceans that covered the United States dried up. It appears in two arrangements: "bedded" or layered, and "domed," as a hill.

Salt has several virtues as a geologic medium. It is abundant, distant from earthquake zones, and inexpensive to excavate. It is almost impermeable as it is plastic—cracks and crevices are sealed by pressure. The committee believed that wastes deposited in salt would be free of contact with water in the future because the very existence of extensive salt deposits indicates past freedom from water.

The favorable opinion of salt led the Atomic Energy Commission to undertake a 10-year investigation named Project Salt Vault. Tests were made in a mine formerly used by the Carey Salt Company near Lyons, Kansas. The mine was cleaned up, repaired, and equipped with a shaft hoist, a ventilating system, and an electric power system. An underground transporter was brought down to the tunnels. Studies included the effect of heat supplied electrically, and the effects of both heat and radiation from spent fuel. Corrosion measurements showed ordinary steel to be much better than stainless steel in a salt environment. The tests were successful, and it was concluded that salt was indeed a good medium. The AEC plan to put a repository in Kansas was canceled in 1970, however, when previous drilling for oil and gas was discovered and that large amounts of water had been pumped in to remove salt.

Despite this experience, salt continued to be a promising disposal medium. The Waste Isolation Pilot Plant (WIPP) is a repository that has been constructed in salt beds some 250 million years old in southeastern New Mexico, 25 miles east of Carlsbad Caverns. Its purpose is the safe disposal of transuranic wastes, of which there were originally some 185,000 cubic meters stored at DOE sites around the country. The whole WIPP site comprises more than 10,000 acres, most of which is buffer. The facility has surface waste-handling facilities on an area of 30 acres. The site is on a bedded salt layer 2000 feet thick, located 850 feet below the surface. Wastes are placed some 2150 feet down, in the middle of the salt layer.

Work started in 1982. Four vertical shafts were drilled down from the surface for air intake, salt handling, waste delivery, and air exhaust. Excavation at the lower level is eventually to produce eight underground "panels," with each panel having seven rooms, 10 m wide and 91 m long.

Environmental data were collected and geological measurements made related to performance assessment. The DOE had planned to start in 1991 to emplace drums of TRU for test purposes, but the project was delayed by legal action related to the withdrawal of New Mexico land for the project. Congressional action in 1992 released the land but added numerous requirements, including the need for certification by the Environmental Protection Agency of the future safety of the facility. Testing took several more years, until 1999, when the facility opened. Meanwhile, containers of TRU were deteriorating, and it was necessary to repackage some of the waste. One of the requirements of TRU disposal at WIPP is accurate knowledge of the contents of waste containers, including materials, radioactive content, and radiation levels. Some difficulty was experienced in establishing representative sample drums for test purposes.

The first truck shipment of TRU in TRUPACT II containers is shown in the photograph on the next page, taken March 26, 1999. In a third-year celebration in 2002, the operating company, Westinghouse TRU Solutions LLC, was honored and note made of 700 shipments and the disposal of a total of 19,000 drums of waste. Over the life of the facility, a total of 37,000 shipments are expected from many sites in the United States. Most will be from national laboratories, Savannah River, and Hanford. The site will be sealed in 2039, after which there will be a 100-year institutional control period. Prominent markers and various barriers will be put in place to warn potential intruders.

DOE has opened a corridor at WIPP for research projects that require lower-than-average radiation levels. The thick salt layer is an excellent shield for most cosmic radiation, and the radioactivity of

TRUPACTII containers arrive at the Waste Isolation Pilot Plant.
(Courtesy of Waste Isolation Pilot Plant.)

the salt is extremely small. Research is conducted on neutrinos from space and the biological effect of low-level radiation. Of scientific biological interest is the discovery of spores of bacteria some 200 million years old in the salt deposit. The facility will also be used for investigations related to waste disposal.

Of special technical interest is the movement of salt brine, which tends to migrate under the influence of pressure and temperature. Studies will also be made of the release of gases as the result of interaction of salt with containers and waste. Continued assessment will be made of the expected long-term (10,000 yr) performance of the proposed repository, from both chemical and radiological standpoints.

Environmental Restoration*

Over the 40 years of the Cold War with the Soviet Union, the primary mission of the Atomic Energy Commission, the Energy Research and Development Administration, and the Department of Energy was weapons production, with less attention to environmental controls. As a consequence, many sites around the country

---

*Appreciation is extended to Victor S. Rezendes of the General Accounting Office for valuable information on this topic.

became contaminated with radioactive materials by the nature of the processes involved. Even before the need for nuclear weapons was reduced by events in the former Soviet Union, it was seen that a major cleanup was necessary. In 1988 Congress included language about environmental contamination in the Defense Authorization Act. Two descriptions of the needs were developed—the Glenn report and the Salgado report. In 1989 DOE Secretary James Watkins initiated a number of actions in response to evident needs. An assessment of compliance with rules and laws was made that revealed that environmental and safety problems existed at many sites: 111 U.S. inactive sites were identified, and a goal of cleanup by the year 2019 was set. To accomplish the 30-year task, DOE created the Office of Environmental Restoration and Waste Management, later Environmental Management. In 1996 estimates were made of the cost of cleanup, with a mid-range value of $227 billion. As many as 10,000 locations—buildings and areas—were found to be involved.

The situation at Rocky Flats illustrates the problems to be encountered. This facility near Denver had been fabricating plutonium weapons parts for many years. To evaporate water contaminated with transuranics, five "solar ponds" were dug. When concern arose about the effect on groundwater, a program of sediment removal was begun. Sludge was mixed with concrete to form blocks of what was called "pondcrete." These were discovered to actually be mixed waste (see Chapter 19). Moreover, blocks crumbled because of improper preparation. That waste had to be repacked, and four more ponds were cleaned out. Material classed as TRU goes to WIPP in New Mexico, and LLW goes to a disposal site in Nevada.

The DOE cleanup task is characterized by the need to maintain a large number of records:

(a) The geographic location in the United States

(b) Amounts of waste in inventory, usually measured by volume in cubic meters and by activity by curies

(c) Types of waste—HLW, LLW, TRU, MLLW, mill tailings, etc.

(d) Physical form—solid, liquid, gas, slurry, or vitrified

(e) Character and condition of container

(f) Chemical composition

(g) Radioisotope composition

(h) Current management—in process, stored, buried (retrievable or nonretrievable)

(i) Treatment and/or repackaging required

(j) Mode of disposal

(k) Priority for action.

At the administrative level, many issues and challenges face the program: (a) Coordination with the EPA Superfund program, which has identified a number of contaminated sites; (b) The extent of cleanup that is feasible or necessary (In other words, "how clean is clean?"); (c) The need for new technologies that will be more effective, safer, and cheaper; (d) An adequate supply of trained personnel; (e) Achievement of understanding and support by the public for a long and expensive operation; (f) Establishment of priorities for cleanup; (g) Deciding on the applicability of dose limits for workers, including the concept of ALARA; (h) Finding sites for disposal facilities that will accommodate all the waste; and (i) Development of an effective management process.

In its attack on the complex defense waste situation, DOE prepared a number of comprehensive planning documents. Some of these are available on the Internet (see References). Among them are: (a) The *1996 Baseline Environmental Management Report*, which lists budgets and schedules for remediation, statements of mission, historical background, and plans for action and future use; (b) *Linking Legacies* (1997), which gives background on waste generation in World War II and the Cold War, with abundant descriptions, photographs, and data; (c) *Accelerating Cleanup: Paths to Closure*, containing flowcharts of planned cleanup over the period to 2070, and activities and costs for cleanup of 353 sites; (d) *From Cleanup to Stewardship* (1999), a companion to the report in (c), emphasizing stewardship. This particular document is important in that it highlights a major policy decision on radioactive waste management.

Ideally, the sites that have contamination with radioactive materials would be cleaned up to an extent that would permit residential or other unrestricted use. The Department of Energy has concluded that this goal cannot be accomplished at many locations, and that long-term stewardship is required. Stewardship implies monitoring, maintenance, and control of materials for the foreseeable future or until decay has rendered the materials innocuous. The basis for the conclusion is the impracticality of remediation or an excessive cost, along with the risk to workers who would process the wastes. One good example is the Nevada Test Site, where many nuclear weapons were exploded. Another is the residue that will be left in Hanford tanks after the bulk of the contents have been removed. Attempts to get the last of the waste out would damage tank walls and increase the possibility of contaminating groundwater.

The policy clearly places responsibilities on future generations, but it can be argued that the dedication of enormous fiscal resources to effect more perfect cleanup would jeopardize future prosperity. Stewardship appears to be a necessity in a world with finite resources and almost unlimited human needs.

Cleanup of contaminated facilities used during the Cold War is being performed in a number of states. The DOE's Office of Environmental Management maintains the Central Internet Database (see References) that has an individual Web site for each of more than 30 states, showing the location of the various installations, describing the recent accomplishments, and indicating future goals.

Disposition of Nuclear Weapons Material

The dramatic change in relations between the United States and the republics of the former Soviet Union had an indirect impact on the waste situation. The signing of the Strategic Arms Reduction Treaties (START) resulted in large surplus stockpiles of highly enriched uranium (HEU) and weapons-grade plutonium.

Arrangements have been made for the United States to purchase HEU from Russia for $12 billion, for dilution in both countries to typical low-enrichment uranium (LEU) by the addition of natural or depleted uranium. The LEU will then be used as fuel for nuclear power plants. The agreement calls for HEU to be processed at a rate of 10,000 kilograms per year for 5 years and then at 30,000 kilograms per year for 15 years. The result will be 15 million kilograms of 4.5 percent LEU. The program called "Megatons to Megawatts" (see References) is administered jointly by the U.S. Enrichment Corporation and the Russian firm TENEX. Some additional wastes will be generated in the conversion facilities. Their nature will depend on the blending process.

Excess U.S. plutonium is extracted from nuclear weapons no longer needed. The disassembly of several thousand U.S. weapons is taking place at the Pantex Plant near Amarillo, Texas. This DOE facility has been used for making the chemical high explosive for nuclear weapons, and for assembling and storing weapons. Warheads and missiles are shipped from military forces to Pantex in special armored vehicles with escort and radio tracking. Disassembly is carried out in cells called "Gravel Gerties," so named for the mass of gravel to absorb energy should a chemical explosion occur. The high explosive is carefully removed from the plutonium "pits," which are stored in special bunkers called "igloos," with massive concrete blocks placed in front of their doors.

The tragic events of 2001 have heightened concern about the possible use of weapons material for terrorist acts against the United States. There is similar concern about theft for sale to a country seeking nuclear weapon capability. Several options for the disposition of excess plutonium are available. One is to continue secure storage, with rigorous accountability to avoid diversion. The fissionable material might eventually be used in the startup and operation of future breeder reactors.

Based on a recommendation by a National Research Council committee (see References), the Department of Energy adopted a two-pronged approach to the plutonium problem. One is to convert the metal pits into plutonium oxide and blend it with natural or depleted uranium oxide to form mixed oxide (MOX). The resultant fuel is then to be used in certain commercial reactors to produce useful power. The spent fuel would be treated the same as the ordinary variety. The other approach is to consider the excess plutonium as a waste. The plutonium would be denatured by "spiking" it with fission products or a highly radioactive isotope such as cesium-137 (30.2 yr). This would provide a significant deterrence to diversion of the weapons material during storage. The mixture would be vitrified in one of the plants at Savannah River or Hanford, and held until shipment to the HLW repository.

The possibility of treating plutonium as a waste is one small facet of a complex problem of international proportions. For a comprehensive review of the situation, see the book by Garwin and Charpak (see References).

A comprehensive study of the options and issues will reveal strong differences in opinion, related to concerns about proliferation of nuclear weapons, the desire to derive economic benefit, and attitudes about nuclear energy in general.

CHAPTER 21

Disposal of Spent Fuel

The rate of accumulation of spent fuel as a by-product of nuclear power generation is easily predicted. Most of the reactors on order in the United States have been completed and are operating. Whenever possible, those reactors will operate at maximum capacity in the interest of economy. The amount of uranium-235 consumed depends largely on the energy produced, and the requirement to remove fuel is based on the consumption of fissile material. Thus, assuming a once-through cycle, one can estimate rather accurately how much spent fuel must be managed in the next few decades. The graph shows this amount up to the year 2020, assuming no premature shutdowns and not taking account of possible new reactor installations.

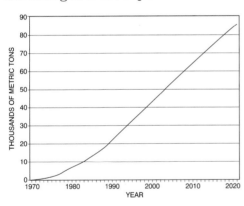

Cumulative U.S. commercial spent fuel discharges, 1970-2020. (Source: DOE Energy Information Administration).

What to do with spent fuel has been an important challenge for the United States since the early 1970s. To answer that question, several others were posed:
- What options are technically possible?
- What strategy is economical at present?
- What alternatives are available long term?

One can readily think of many ways to manage spent fuel and other high-level wastes:
- (a) Dilute and disperse in the air and large bodies of water. This is an unacceptable method for evident environmental reasons.
- (b) Delay action. This approach is favored by those who believe that the present political climate makes disposal anywhere

difficult and who expect public opinion about radioactive waste to become less negative in the future.

(c) Store for future use. Many in the nuclear community feel that as conventional energy sources are depleted in the twenty-first century, materials in spent fuel will be needed for breeder reactors.

(d) Reprocess, retrieve, and recycle. Fissionable elements are especially useful, while some of the radioactive fission products have economic value.

(e) Irradiate waste with neutrons to change the composition and the half-lives of key radioisotopes.

(f) Isolate spent fuel. This is the current preferred method to prevent radioactivity and radiation from affecting living beings and the environment.

The following paragraphs explore the various processes for implementing some of the options.

Reprocessing and Fractionation

In the early days of nuclear power, it was planned to reprocess spent fuel to separate fission products from uranium and plutonium, which would be recycled. Circumstances described in Chapter 14 currently prevent that option from being exercised in the United States. The economic benefit of reprocessing is marginal at best. No large-scale reprocessing facilities are available. More importantly, in the existing political and economic uncertainty set by public opinion, the industry has little enthusiasm for developing, licensing, and constructing a reprocessing plant.

It is of interest, however, to examine some benefits of reprocessing and of fractionation, which is another word for "separating." Separation to prepare for future treatment or use is called "partitioning." First, let us look at the composition of residues from reprocessing. For each metric ton (1000 kg) of spent fuel the weights in the high-level residue are as listed below. Note, the weight of reprocessed

Composition of Residues from Reprocessing

Material	Weight, kg
Fission products	28.8
Fuel	
Uranium	4.8
Plutonium	0.04
Transuranics	
Neptunium	0.48
Americium	0.14
Curium	0.04
Reprocessing chemicals	68.5
Total	102.8

waste material is one-tenth the weight of spent fuel. Of course, as most of the radioactivity is from fission products, the activity per gram of waste is considerably higher.

If the fuel is going to be chemically processed to isolate five elements, it would be prudent to go further and separate out additional groups or individual elements. Recall that strontium-90 and cesium-137 are the worst offending isotopes during the first few hundred years of waste storage. If they were extracted, the residual wastes would contain only a thousandth as much activity. The volume of those radioisotopes is small, and storage could easily be managed. On the other hand, capsules of cesium-137 with its 0.66 MeV gamma ray would be very valuable to the future food irradiation industry as a substitute for cobalt-60, which must be made in a reactor and which uses up neutrons. The B Plant at Hanford, used for separating plutonium in World War II, was adapted for the recovery of cesium and strontium from liquid wastes in the period 1974-1985. A total of 1315 cesium capsules and 601 strontium capsules are now in storage.

By means of existing chemical technology or an extension based on research, there is a possibility of isolating some of the radionuclides of very long half-life that dominate the performance of a repository. Examples are technetium-99 (213,000 yr), neptunium-237 (2.14 million yr), and iodine-129 (17 million yr). Removal of both the shorter-lived heat producers and longer-lived hazardous isotopes might make the disposal process simpler, cheaper, safer, and more acceptable politically. These long-lived isotopes could be stored separately, with only minimum shielding required.

Objections can be raised to the operation of plants for reprocessing and fractionation of spent fuel, because it might make plutonium more accessible for illegal purposes, including the manufacture of nuclear weapons or its use in terrorism. In contrast, one can argue that the best thing to do with plutonium is to place it in a reactor core, which is intensely radioactive already, and burn it there to help produce electric energy.

Another reason for not treating spent fuel is the increase in radiation exposure for workers required to operate, maintain, and repair a reprocessing plant. There is, however, a good possibility of developing remotely maintained facilities, especially if new electronic monitoring and control equipment is coupled to robotic technology. There is an analogy with future space exploration, in which human beings may not be exposed to risk if the missions can be accomplished by electro-mechanical robots instead.

Transmutation

The word "transmute" means to convert one element into another by nuclear reaction. It is like the goal of the alchemist of the Middle

Ages—to turn base metal into gold. Such a process is now possible through neutron bombardment, at a great cost, of course.

As applied to nuclear wastes, transmutation would involve irradiation of wastes by neutrons as in a fission reactor or in a future fusion reactor.

The neutrons are absorbed to produce new isotopes that may have a short half-life or be stable. The process aids natural decay by effectively shortening half-lives.

The elimination of cesium and strontium is difficult because of the high neutron streams required. More likely targets are the transuranic elements plutonium, neptunium, americium, curium, and californium. Other candidates for transmutation are iodine-129 and technetium-99. The elements would be partitioned in a special reprocessing system, fabricated into fuel, and used in a reactor. Because some nuclei are fissile, they would yield power. Disadvantages include the possibility of creating heavy isotopes that decay with undesirable isotopes in the chain. Extra protection would be required in fuel fabrication because of neutron emission and residual radioactivity.

The method appears to be very attractive, but it is currently not feasible on the large scale needed. If research and development were resumed on the Integral Fast Reactor (Chapter 14), burning of transuranic radionuclides could become a reality. The on-site fuel reprocessing facility that accompanies the reactor would make it safe and convenient to recycle certain isotopes.

Exotic Disposal Concepts

Among the various ideas advanced for disposing of high-level wastes were three, labeled space, ice-sheet, and seabed.

Space disposal. This would involve carrying packages of waste by shuttle to place them in orbits around the earth or sun, as new "asteroids." The advantage is obvious, but drawbacks include (a) the possibility of an aborted mission, with burnup during reentry and contamination of the atmosphere or (b) excessive cost because of the large payloads to be transported.

Ice-sheet disposal. In this scheme, canisters of waste would be taken to Antarctica and laid on the surface of the ice. Residual fission product heat would cause the containers to melt their way down to the solid rock. Objections include (a) the very brief window of time for access and (b) the possibility that waste would escape to the sea in a water layer between ice and rock.

Seabed disposal. Here, canisters would be taken by ship out to sea. They would have a sharp point and fins to guide them point down into the sediment at the bottom, where they would bury themselves. Eventually, the dissolved waste would be diluted in the vast volume of water. This technique has not been pursued because of uncertainty about safety.

Geologic Disposal

Geologic disposal simply means burial in the earth. Several ways have been considered. First is placement of wastes in a very deep hole, going down 6 to 10 miles beneath the surface, and well below the level of moving groundwater. Canisters would be lowered into the hole and stacked in a column some miles in height, with the hole plugged. The advantage is remoteness; the disadvantage is lack of experience with large holes at such depths. Also, it is difficult to learn geologic features at such depths.

Second is a rock-melting technique, in which wastes would be dropped into a hole a mile deep. As visualized, the heat from radioactive decay melts the rock at the bottom, the wastes mix with the rock, and the mass of rock-waste eventually cools and solidifies, after perhaps as long as a thousand years.

Third is a process called "hydrofracture," in which water is forced into rock such as shale, causing layers to separate. Then waste mixed with cement or clay is pumped into the spaces, where it solidifies. The method has been used in the United States and the former Soviet Union, but it requires very complete knowledge of the geology.

Fourth is the island isolation method. Waste containers would be placed below the fresh water table on an uninhabited small island far from civilization. The concept lies between seabed disposal and land disposal. It is not high on the priority list.

Fifth, and the preferred geologic method, is use of a mined cavity. As originally conceived in the early 1970s, a shaft would extend from the earth's surface down to a series of horizontal tunnels in a geologic medium, and canisters containing spent fuel would be inserted in holes drilled in the corridor floor. That method of emplacement of wastes has since been modified, but the use of underground disposal has been adopted. The advantages are (a) conventional mining techniques can be applied, (b) the waste can be retrieved if necessary or desirable, and (c) the waste is well isolated from the environment, allowing for decay of most of the isotopes. With suitable spacing of waste containers, problems of heat generation can be minimized.

Choices of Geologic Medium

Several materials have been promising candidates for the host geologic medium for a U.S. high-level waste repository:

- Rock salt, as recommended by scientific bodies and as used in the WIPP site (Chapter 20). In addition to New Mexico, salt is found in tall underground columns called salt domes in Texas, Louisiana, and Mississippi. In the 1970s plans were laid by the

AEC for the disposal of wastes in salt near Lyons, Kansas. The discovery of numerous prior drillings forced the abandonment of that project.

- Granite, prevalent along the Appalachian chain and the Great Lakes region. Earlier plans for a site in the East had considered this form of crystalline rock.

- Basalt, a rock formed by lava from volcanoes, found in the Northwest near the Columbia River.

- Tuff, a compacted and hardened ash from volcanic action millions of years ago. It is found in the Southwestern United States, in particular at Yucca Mountain in Nevada, the likely site of the future high-level waste repository.

All of the above media were considered in the Nuclear Waste Policy Act of 1982, passed by Congress (see Chapter 22). The ability of these geological media to greatly retard the migration of wastes in water, allowing decay to occur, depends on physical and chemical properties. One is porosity, which is the fraction that is space. Pore size and connection are important: small pores only slightly connected retard migration best. Second is permeability, a number that is related to ease of water flow. Permeability values are listed for several materials, all with a porosity of about one-half:

Permeabilities of Geologic Media

Material	Permeability
Clay	0.000016
Silt	0.33
Sand	25.3
Gravel	1130

Roughly, these numbers represent the rate of water flow under a certain water pressure. If the rock is cracked, the flow may be higher than if it is uniform.

The interaction between the radioactive contaminant and the geologic medium is expressed by a quantity labeled K_d, called the distribution coefficient of the chemical reaction. If it is large, the waste moves very slowly; if it is small, the waste moves with the same speed as the water. The retardation factor R, which is the ratio of water speed to contaminant speed, depends on K_d, density ρ, and porosity ϕ through the formula $R = 1 + K_d \, \rho/\phi$. Note that R goes to 1 for a very porous medium and increases with the reaction constant.

Approximate values of R for selected elements and media are listed in the table on the next page.

The wide variation of R for the different elements means that each radioisotope must be considered separately in the analysis and calculation of system performance.

Retardation of Elements in Geologic Media

(From *A Study of the Isolation System for Geological Disposal of Radioactive Wastes*, National Academy Press, Washington, D.C., 1983, page 147.)

Element	Medium	Retardation Factor
Iodine	All	1
Technetium	All	5
Cesium	Salt	10
Neptunium	Tuff	100
Plutonium	Basalt	500
Thorium	Granite	5000

Characterization of a Repository

A high-level radioactive waste repository must protect people and the environment for many centuries. Any transport of material must result in a radiation dose that is less than a regulatory limit such as 15 millirems per year. To ensure this objective, the system, consisting of waste form, containers, backfill, and geologic medium, must minimize water access and contaminant escape. The waste must be emplaced in a setting that is safe initially and is likely to remain so for a very long time.

To achieve confidence in the security of the waste, comprehensive site characterization studies are made. The investigation is accompanied by analyses that predict the radiation exposure over time. A description of the action plans for the Yucca Mountain Project (YMP) will give us perspective on the magnitude and complexity of a full characterization process. The plans were designed to lead to a decision as to whether the site is suitable. If it is, commercial spent fuel and vitrified defense wastes will be emplaced there.

Yucca Mountain, seen in the photograph on the next page, is in Nevada, about 100 miles northwest of Las Vegas, and very near the Nevada Test Site for nuclear weapons. The water table is about 1800 feet below the surface. It is believed that heat from decaying fission products would dry the rock surrounding the waste containers. Most of the required studies relate to the unsaturated zone, above the water table.

Many sciences, popularly called "ologies," were applied: geology—the nature of rock; hydrology—the flow of water; seismology—earthquake effects; volcanology—potential for eruption of volcanoes; meteorology—climate and weather patterns; ecology—the science of biological systems—as well as law, sociology, demography, and politics. Some of the sciences, such as geology and hydrology, are closely related to each other; for example, fractures in rock affect the rate of water flow. Involved in the Yucca Mountain investigation were hundreds of scientists from many organizations. Included were the Department of Energy, the U.S. Geological Survey, and the national

Yucca Mountain, the potential site of a waste repository. This aerial view of the northeast shows the Exploratory Studies Facility. (Courtesy of the Yucca Mountain Site Characterization Project Office, U.S. Department of Energy.)

laboratories—Sandia, Los Alamos, and Lawrence Livermore—as well as several companies.

The testing program consisted of drilling deep holes to find out how the geology varies in three directions; cutting trenches to investigate geologic faults, in which changes could cause earthquakes; measuring water presence, chemical composition, and flow patterns; and determining the presence of valuable mineral deposits that might attract intruders. To focus on the needs for information, the YMP set up an Exploratory Studies Facility. This is an underground laboratory consisting of ramps down from the surface to a series of tunnels. A number of boreholes were drilled using a minimum of water and using dust suppression systems to avoid air pollution. Several trenches were dug to look for geologic faults and to make predictions of seismic activity.

The above physical studies were supplemented by observations and data collection on environmental and societal aspects of the region. Animal and plant species were identified and tabulated, with special attention to those classified as protected, threatened, or endangered. Desert tortoises, for instance, are rare and threatened, while lizards, ground squirrels, and scarab beetles are common and unprotected. Information on factors that would affect the small human population in the area were also gathered.

The characterization process took about 15 years to complete. Its objective was to assess the suitability of Yucca Mountain as the site for a repository. A large number of individual qualifying and disqualifying criteria were applied. A few examples were:

Qualifying:
- environmental quality can be protected
- adequate transportation routes are available
- societal effects can be minimized
- land acquisition is possible

Disqualifying:
- population density nearby is high
- site is in a federally protected area
- significant pathways from previous mining are present.

In 2002 the Secretary of Energy recommended the site to the President in a document accessible at various locations (see References). The Secretary reviewed the history of the project, the extensive site characterization studies, and the criteria used to make a recommendation. The suitability evaluations made by DOE were of two types. One is pre-closure, the period of 50 to 300 years during which casks of waste are put in place. This involved consideration of conventional radiological safety. The other is post-closure, the period up to 10,000 years, which required analysis of possible scenarios.

The role of water is taken as an example. Yucca Mountain has low rainfall and less than half an inch of water would reach the repository, 800 feet down through rock, which represents the natural barrier system. The engineered barrier system consists of multiple layers of protection and a physical arrangement that minimizes contact of water with wastes. The barrier system is evolving as new information is available, but the current concept uses features as diagramed.

Engineered Barrier System

Spent fuel disposal at Yucca Mountain.

The "drift" is a cylindrical cavity cut in the rock, into which the packages of spent fuel or vitrified waste are placed. On the inside of the waste package is a cask with a gridded basket to hold spent fuel assemblies. Layers of metal around the basket are designed to last thousands of years. The outer 1-inch-thick barrier is nickel-based Alloy 22; the inner 2-inch-thick barrier is 316 stainless steel. Above each waste package is a drip shield that acts as umbrella, diverting water from the container. It is composed of titanium with Alloy 22 supports. A V-shaped trough beneath the waste package permits water to drain away. The whole assembly is emplaced in the drift by use of a rail system. Even if some dissolution occurs, the contaminated water has another 800 feet to go down to the water table.

Calculations using the Total System Performance Assessment, using the characteristics of both the natural and engineered barriers, predict maximum doses to individuals of the public of 0.1 mrem/yr, in contrast with the regulatory limit of 15 mrem/yr and the average U.S. radiation exposure of 360 mrem/yr. The recommendation by the Secretary of Energy comments on the amount of information provided to the public and on the issue of transportation of spent fuel and other wastes. The summary contains the Secretary's belief that moving forward towards a repository at Yucca Mountain is in the national interest.

The President accepted the recommendation. Congress overrode an objection by the state of Nevada and authorized DOE to develop the site. The first step would be the initiation of an application to the NRC for a construction license around 2004. Completion of the facility and the start of shipments of spent fuel from utilities are expected around 2010.

The next two chapters discuss the laws and regulations that govern waste management, and the societal aspects of the disposal of all types of radioactive waste.

CHAPTER 22

Laws, Regulations, and Programs

Laws and Regulations on Atomic Energy

The management of radioactive materials has been subject to laws and regulations for over 50 years. Requirements have evolved as experience has been gained, and programs have changed with the political climate. To review the historical background will help us understand the current situation.

The Atomic Energy Act of 1946 established the Atomic Energy Commission (AEC) and directed it to conduct research and development on peaceful applications of fissionable and radioactive materials. The Atomic Energy Act of 1954 emphasized both domestic and international uses of the atom. It also provided for the control of "source material" (uranium and thorium), "special nuclear material" (plutonium and enriched uranium), and "by-product material" (other radioactive substances such as fission products).

The National Environmental Policy Act of 1969 (NEPA) has as its purpose "to prevent or eliminate damage to the environment and biosphere and stimulate the health and welfare of man." The act created the Council on Environmental Quality (CEQ), an advisory and coordinating group reporting to the President. NEPA requires every federal action that may significantly affect the environment to be accompanied by an environmental impact statement (EIS). An EIS is a large document that describes alternatives and associated potential environmental, social, and economic effects; includes public comments and agency answers; and reports the findings of hearing boards. The Environmental Protection Agency (EPA) was created, with responsibility for air and water standards, limits on pollutants, and control of radioactivity. The EPA provides for public participation through meetings, hearings, and advisory group reviews.

The Energy Reorganization Act of 1974 divided the function of the AEC—developmental and regulatory—between two agencies, the Energy Research and Development Administration (ERDA) and the Nuclear Regulatory Commission (NRC). The Arab oil boycott of 1974 caused the United States to restructure its energy plans. The Energy Organization Act of 1977 replaced ERDA by the current Department of Energy (DOE), which also absorbed several other energy-related government programs.

The NRC has jurisdiction over reactor construction and operation. It also licenses and regulates the possession, use, transportation, handling, and disposal of radioactive materials, including wastes. Within NRC guidelines, Agreement States accept authority to control radioactive materials. The principal reference is the *Code of Federal Regulations Title 10 Energy*. For example, 10 CFR 20 covers radiation standards, and 10 CFR 50 covers reactors. Most people do not realize how highly regulated nuclear energy activities are. Title 10 Energy contains about 1400 pages of rules. The recurring theme is assurance of health and safety of the public and protection of the environment.

Other important agencies are the Department of Transportation (DOT), which provides rules on the shipment of all kinds of radioactive materials, and the Federal Emergency Management Agency (FEMA), which has plans for response to emergencies that involve the release of radioactivity.

The relationship of the three agencies that most directly affect waste management can be summarized as follows: EPA provides radiation protection standards; NRC licenses and regulates waste disposal under EPA limits; DOE does research and development, advises and helps with low-level waste disposal programs, and will operate defense and high-level commercial waste repositories.

Major legislation in the areas of LLW (1980, 1985), HLW (1982, 1987), TRU (1992), and energy (1992) provides the basis for the U.S. programs in waste disposal. The sequence of laws reflects responses to problems and to changes in public opinion. Regulations implementing the laws continue to evolve, affected by court decisions. There is a complex interaction among Congress, the states, regulating agencies, industry, and the judicial system. The situation is briefly reviewed in the following sections.

The Low-Level Waste Policy Acts

Concern about the possibility of having no place to dispose of low-level radioactive wastes led the states to seek control of waste management. This led to the enactment in 1980 of the Low-Level Radioactive Waste Policy Act. It says, "Each state is responsible for providing for the availability of capacity either within or outside the state for disposal of low-level radioactive waste generated within its borders (excluding defense or other federal wastes). Low-level radioactive waste can be most safely and efficiently managed on a regional basis." The act goes on to say that the states may enter into compacts with their neighbors under congressional authorization. The law called for the ability of compacts to exclude wastes from other regions after January 1, 1986. The legislation led to the formation of a number of compacts, including the Appalachian, Central Midwest, Central States, Midwest, Northwest, Rocky Mountain,

Southeast, and Southwestern, with the newer Texas (with Vermont and Maine) and Atlantic. The map shows the compacts as of the end of 2000. Compacts are responsible for deciding what facilities are needed and which state will serve as host and for how long.

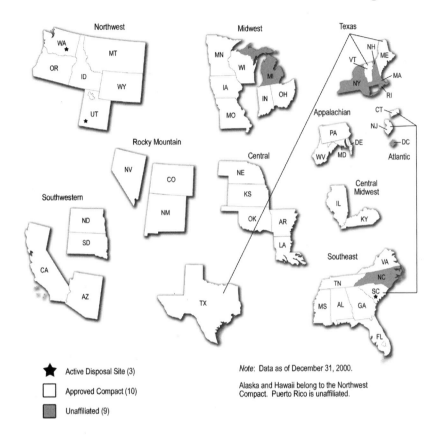

Interstate compacts for low-level waste. (Source: Nuclear Regulatory Commission.)

Negotiations among many states to form compacts and start developing disposal sites took longer than expected, making it impossible to meet the 1986 deadline. To relieve this problem and at the same time to expedite action, Congress passed the Low-Level Radioactive Waste Policy Amendments Act of 1985 (LLRWPAA). It called for keeping the three commercial disposal sites open for use by all through 1992. Annual and total limits on the volume of waste that can be sent from reactors were specified. The LLRWPAA called for milestones and deadlines on ratifying compacts, selecting host states, developing plans, submitting license applications, and providing for disposal. Existing disposal sites were allowed to impose surcharges for disposal of wastes from regions without sites, with rebates to be used by states or compacts for site development. The Department of Energy kept track of these arrangements, with authority to allocate additional emergency disposal capacity to reactors, while the Nuclear Regulatory Commission could authorize emergency access to existing sites. The law also says that DOE is responsible for disposal of commercial LLW that is greater than Class C limits. The Department of Energy is required to assist states and compacts to achieve disposal facilities. To do this, DOE established the National Low-Level Waste Management Program, which provided technical expertise on waste disposal. It also administered the Host State Technical Coordinating Committee and the Low Level Radioactive Waste Forum, both of which ensured communication and coordination among the various compacts that were developing disposal facilities.

Progress towards establishing low-level waste disposal facilities by 1996 was very slow. The task turned out to be much more complex and time-consuming than ever anticipated. A number of reasons for the delay can be cited: (a) each project must develop its own criteria and procedures for site selection; (b) a survey of a complete host state for potential sites involves the collection and analysis of large amounts of data; (c) the processes of site characterization and interpretation of data are long and involved; (d) the application for a license is a many-volume document; (e) the regulatory review of the application is thorough and extensive; and (f) lawsuits initiated by potential host communities or intervening groups delay action, sometimes by years.

In spite of the best efforts of the supporting programs and those of the states and compacts, no new disposal facilities in the United States were built. The reasons for such failure have been identified by the General Accounting Office (see References). One is opposition by members of the public, with support for that opposition by political leaders at all levels. Another is the high cost of developing facilities that meet all regulatory expectations. The declining volume of LLW led to a lack of urgency and encouraged postponement or cancellation of plans. Finally, for the volumes generated, the three existing facilities have been adequate.

Only three commercial low-level disposal sites were in operation in the United States as of 2002. First was the Hanford, Washington facility, which serves the Northwest Compact and the Rocky Mountain Compact (see map on page 177). Second is Barnwell, South Carolina, which accepts waste from all other states but North Carolina. It is planned to limit access by 2008 to the Atlantic Compact, including South Carolina, Connecticut, and New Jersey. Third is Envirocare, in Clive, Utah, where the disposal facility is independent of the compact system. It accepts A-type waste from all states (see References).

When the Low-Level Waste Policy Act of 1980 was passed, Hanford and Barnwell were in existence and were absorbed within the compact system. Eventually, when Barnwell limits access, a new low-level waste management system must be available. Alternatives that would have to originate in Congress are to allow private industry to develop disposal capability, or to assign the Department of Energy the responsibility of accepting wastes in its disposal sites. Prominent among these is the Nevada Test Site where there already is contamination from weapons research. As discussed by Telfer (see References) this 1375 square mile tract is being used for DOE LLW. Waste is placed in subsidence craters from underground tests. However, some of these approaches might not be successful because they do not address the reluctance of communities and states to serve as hosts.

The lack of disposal facilities has forced generators to store waste on site at considerable expense. In some cases, when storage costs or disposal costs were prohibitive, the use of radioactive materials was abandoned, to the detriment of the organization and its beneficiaries.

A Supreme Court ruling invalidated the part of the 1985 Amendments Act stipulating that states must take title to wastes if a facility was not available by 1996. The rest of the law remained intact.

The Nuclear Waste Policy Acts

For the management of high-level radioactive waste and spent fuel, the Nuclear Waste Policy Act of 1982 was passed by Congress. It was a compromise among industry, government, and environmentalists, containing timetables for action by the Department of Energy leading to underground disposal of HLW.

The act relates primarily to commercially generated materials but provides for the disposal of defense wastes upon Presidential approval. The Nuclear Waste Fund was set up to pay for disposal, with money coming from the waste generators, who in turn charge users of electricity. A fee of one-tenth a cent for each kilowatt-hour is assessed. This is to be compared with a typical cost to the consumer of 6 cents per kilowatt-hour. Although the Nuclear Waste Fund is accumulating, it is necessary for Congress to appropriate the money for operations.

Following the dictates of the law, DOE set up an Office of Civilian Radioactive Waste Management (OCRWM), with its director reporting to the Secretary of Energy. Guidelines were issued for the process to select suitable sites for a repository, a mission plan was developed, and geological surveys were begun. Several sites were identified as potentially acceptable for the first repository: in basalt at Hanford, Washington; bedded salt in the Paradox Basin in Utah and the Palo Duro Basin in Texas; salt domes in Mississippi and Louisiana; and tuff at Yucca Mountain in southern Nevada. At the same time, possible locations were investigated for a second repository, in the East, in crystalline rock. The Great Lakes area and the Appalachian range were principal candidate regions.

The use of a Monitored Retrievable Storage facility (MRS) was studied, in accord with the law. Congress had visualized the MRS as an alternative to a repository, but the Civilian Waste Management Program conceived it as one part of an integrated disposal system. Fuel assemblies would be shipped from reactor storage pools to the MRS as a staging area, where fuel would be packaged and shipped to the disposal sites. The MRS could also provide backup storage capacity in case the opening of a repository was delayed.

In 1987, the choice of sites was narrowed to Hanford, Washington; Yucca Mountain, Nevada; and Deaf Smith County, Texas; and site characterization studies were begun. The search for repository sites and the MRS created much concern among citizens and lawmakers. The decision by DOE to suspend the search for an eastern site was questioned. After much political compromise, Congress passed the Nuclear Waste Policy Amendments Act of 1987, which restructured DOE's HLW program. The only western site to be characterized would be Yucca Mountain. Nevada would receive financial compensation and special consideration in federal research projects.

The status of the MRS was redefined. A study commission would evaluate the need for it, and limits on fuel storage capacity of the facility are set at 10,000 tons. The NRC must issue a repository construction license before the MRS can be built; the license would be the legal device that would prevent the MRS from becoming a permanent storage facility. A Nuclear Waste Negotiator position was created, with assignment to seek a volunteer host for the MRS. Interest was expressed by some Indian tribes, but nothing came of it. The concept may be implemented at the Nevada Test Site.

The 1987 act had a number of special features. A Nuclear Waste Review Board in the National Academy of Sciences was created; spent fuel must be shipped in NRC-approved packages, with state and local authorities notified of shipments; authority was given for continued study of the subseabed disposal option, a topic of interest to European countries. No further crystalline rock studies were allowed, and DOE was to submit in the period 2007 to 2010 a study on the need for a second repository. The decision to characterize only one site would save considerable expense unless the Nevada site is found unsuitable, requiring other locations to be considered. The redirection of DOE's HLW program involved the preparation of a new limited scope mission plan and the release of only one site characterization plan.

Progress in the characterization of the Yucca Mountain site was slow, largely because of efforts by the state of Nevada to stop the project. Questions and concerns about the suitability of the site were raised. The principal issue was whether there were pathways for rapid transfer of radionuclides to the environment. After several years of litigation, DOE obtained approval to proceed and not to be held up by requirements for permits.

Energy Policy Act of 1992

This legislation was long in coming and broad in its scope, emphasizing energy efficiency, research and development on conventional fuels, alternative fuels, and uranium enrichment. The little that was said about radioactive wastes was important, however. The law specified that (a) EPA would set standards for Yucca Mountain based on findings by the National Academy of Sciences on several specific issues related to radiological protection, (b) NRC would provide requirements and criteria based on EPA standards, assuming engineered barriers and long-term oversight of the repository by DOE, (c) DOE (with NRC and EPA) would report to Congress on the adequacy of plans for disposal of waste from future reactors, and (d) states would have authority over below-regulatory-concern (BRC) wastes, negating NRC policy. (The NRC had hoped to exclude wastes that led to doses less than 1 mrem/yr, allowing them to be disposed of as ordinary waste. After failure to achieve agreement among interested parties, the NRC planned to permit disposal of BRC on a case-by-case basis. It is likely that most BRC will be disposed as LLW.)

Environmental Standards

The technical basis of licensing and regulation of potentially harmful substances is normally provided by the Environmental Protection Agency (EPA). Rules appear in the *Code of Federal Regulations Title 40 Environment*. In the case of radiation protection, standards were delayed, and NRC initiated the rules. Also, some EPA rules have been challenged by the courts, requiring new rulemaking. Thus some EPA rules are in place while others are being developed.

The approach to standards taken by the EPA is to look at the possible hazards under a variety of situations ("scenarios") involving different sources of radionuclides, their method of treatment, protective measures taken, and health consequences. The dollar cost of each case is also calculated.

The evolution of EPA rules is difficult to describe because of the interaction with Congress, other agencies, and the public. Only a few highlights are given here. In Title 40 of the *Code of Federal Regulations* there are several parts in Radiation Protection Programs. The first of these is 40 CFR 190, referring to nuclear power operations and limits on doses and releases in the uranium fuel cycle. The second is 40 CFR 191 on management and disposal of spent nuclear fuel, high-level, and transuranic wastes. The third is 40 CFR 197, specific to Yucca Mountain. The fourth part is 40 CFR 192, having to do with uranium and thorium mill tailings. An example is the limit of radium-226 to 5 picocuries per gram in the top layer of soil.

In locating disposal facilities, wetlands must be avoided. Wetlands are defined as bodies of water or land saturated with water that support plant and animal life, excluding farmland. They may be filled in only with special permit. The EPA and the U.S. Army Corps of Engineers administer the wetlands rules of Section 404 of the 1972 Clean Water Act.

The protection of wildlife is also under the EPA. Included are rare, threatened, or endangered species of animals and plants.

Regulations on Low-Level Radioactive Wastes

Rules on civilian radioactive wastes in general are provided by the Nuclear Regulatory Commission. They are based on research by the NRC and its contractors. A draft rule with much supporting information is issued for review by those who would be affected, including industry and the public. After refinement, the rule is published officially in the *Federal Register* and in the *Code of Federal Regulations Title 10 Energy*. The principal regulation for low-level wastes is Part 61, usually called 10 CFR 61.

There is a wide range of activities in the low-level wastes to deal with, from barely above background to those comparable to high-level wastes. The lowest category is "exempt wastes," which can be disposed of without regard to radioactivity. Next are Class A wastes, which require "minimum" precautions for disposal. This means no use of cardboard containers, a need for liquid waste to be solidified or mixed with an absorbent so that there is no more than 1 percent liquid, no explosive or spontaneously combustible material, at limited pressure if gaseous, and treated if of biological origin. Class B wastes must meet minimum requirements but also have "stability." This means they must keep their size and shape despite effects on containers from soil weight, moisture, or radiation. Class C wastes should be isolated from a future "inadvertent intruder," a person who accidentally comes upon the waste while digging in the area after the site has been closed. The person may be drilling a well or excavating for a building or cultivating the land. The C waste should be buried more deeply to protect the intruder. Wastes that are more active than C cannot be given near-surface disposal. These greater-than-Class-C (GTCC) wastes must be treated as HLW, to be disposed of by the Department of Energy.

The boundaries between classes of wastes depend both on the isotope's half-life and the specific activity in curies per cubic meter. For example, when the concentration of tritium (12.3 yr) reaches 40 Ci/m^3, it becomes B waste; when that of cesium-137 (30 yr) reaches 4600 Ci/m^3, it becomes GTCC; when iodine-129 (17 million yr) reaches 0.008 Ci/m^3, it goes from A to C. The regulation 10 CFR 61 gives details for all radionuclides.

The waste disposal facility is licensed either by a state or by the NRC for use by a commercial operator. The site is selected from several candidates on land owned by the state or federal government. All pertinent facts about the geology, water flow patterns, and nearby population must be known. Operations are inspected to be sure that the wastes are properly managed, and if so the license is renewed. At the end of the useful period of the facility, 20 to 30 years, the site is closed. The license is transferred to the state or federal agency, which will continue to monitor the site for the period of institutional control, which is 100 years. Then the license is terminated and no further maintenance is needed, but the design should have assured protection for a period of 500 years.

Supplementing the regulations are NRC documents called Regulatory Guides (colloquially "Reg Guides"), providing information on such things as quality assurance, design bases, calculation methods, and the form for reporting. An example is No. 4.18, "Standard Format and Content of Environmental Reports for Near-Surface Disposal of Radioactive Waste," 1983.

The NRC relies on its Advisory Committee on Nuclear Waste for input on regulations for all forms of waste, on transportation issues, and Yucca Mountain.

Regulations on High-Level Radioactive Wastes

The history of regulation of high-level waste disposal is convoluted and not easy to understand. Each of the federal agencies that were involved—NRC, EPA, and DOE—had developed regulatory material, but each had to make significant modifications for various reasons, including new laws, public input, lawsuits, and differences in philosophy. One can find some of the details of the history on the Internet, especially in the final rules in the *Federal Register* (see References). These documents provide background information, the thinking behind new regulations, and responses to comments by the public.

In response to the Nuclear Waste Policy Act of 1982, the NRC developed the regulation "Disposal of High-Level Radioactive Wastes in Geological Repositories." It was Part 60 of *Code of Federal Regulations, Title 10 Energy*, abbreviated as 10 CFR 60. It was regarded as a generic regulation, applicable to any of the potential sites. Some of its key provisions were (a) the design and operation should not pose an unreasonable risk, (b) multiple barriers are to include the waste form, the container, and the host rock, (c) a thorough site characterization is needed, (d) the HLW is to be retrievable for up to 50 years, and (e) predictions of safety should be conservative and account for uncertainties. Certain performance requirements for the waste packages and the facility were stated. Since these had to be modified to conform to EPA rules, they are now irrelevant, even though regulation 10 CFR 60 is still on the books.

When Yucca Mountain was designated as the principal potential site, NRC was prepared to use 10 CFR 60 to review an application by DOE for a license. However, the Environmental Protection Agency issued the document "Public Health and Environmental Radiation Protection Standards for Yucca Mountain, Nevada," 40 CFR 197. The limits therein on performance differed considerably from those of NRC. For example, the regulatory period was changed from 1000 years to 10,000 years; the annual committed effective dose (see Chapter 6) was reduced from 25 mrem to 15 mrem; a new drinking water standard for beta-gamma emitters was set at 4 mrem/yr.

As a consequence, NRC issued a new revised regulation, "Disposal of High-Level Radioactive Wastes in a Proposed Geological Repository at Yucca Mountain, Nevada," 10 CFR 63 (see References). It incorporated the EPA standards, almost verbatim. Among other new features were (a) reference biosphere, which specifies the environment of potentially affected individuals, (b) a definition of the reasonably maximally exposed individual, e.g., life style, drinking water intake, and diet of adult neighbors to the repository, (c) use of "reasonable expectation" (EPA) instead of "reasonable assurance" (NRC) in relation to protection, (d) exclusion of unlikely features, events, and processes (defined later in terms of a numerical chance of occurrence within 10,000 yr). Regulation 10 CFR 63 is also available on the NRC Web site (see References).

Licensing of Yucca Mountain

The application to NRC by DOE for a license to construct and operate the Yucca Mountain repository will be based on the findings of DOE's investigations on site suitability. The applicable regulation is "General Guidelines for the Recommendation of Sites for Nuclear Waste Repositories: Yucca Mountain Suitability Guidelines." Some of the key features of 10 CFR 963 are:

(a) emphasis on total system performance assessment, which identifies features, events, and processes that might affect public radiation dose, considering probabilities and significances. Engineered and natural systems are to be analyzed together.

(b) distinction between pre-closure and post-closure safety,

(c) catalog of suitability categories:

(1) ordinary processes with nine criteria

(2) disruptive processes and events—volcanism, seismic events, nuclear criticality, and inadvertent human intrusion.

In carrying out the major task of transporting spent fuel across the country, the DOE will have to follow regulations of the Department of Transportation. Relevant parts are in the range 49 CFR 171-179, which cover all types of hazardous material. Rules are found on records, packaging, labeling, and special features of the modes of transport—rail, truck, and air. The regulations are available on the Web (see References) with the caveat that each of the hundreds of subparts uses the same filename.

Finally, DOE will need to comply with those special regulations of the Environmental Protection Agency that are not contained in NRC regulations. Examples are (a) care to avoid harming endangered species, such as the desert tortoise, including moving them to other locations, (b) assuring protection against creation of dust during construction operations, and (c) maintenance of information on the socioeconomic impact of new projects as required by federal law.

The diagram suggests the relationship of the four main cognizant federal agencies: NRC, DOT, EPA, and DOE. Considerable effort has been expended to ensure compatibility among the regulations. In addition, the Nuclear Waste Technical Review Board oversees DOE's work on Yucca Mountain and makes recommendations (see References).

Nuclear Regulatory Commission
10 CFR 63
Enforces standards; licenses and inspects

Department of Energy
10 CFR 963 Designs, constructs, and operates repository

Department of Transportation
49 CFR 171-179
Regulates transport of waste

Environmental Protection Agency
40 CFR 197
Establishes radiation standards

Relationship of federal agencies on regulations for the Yucca Mountain high-level waste repository. (Adapted from NRC Web page by Nancy Reid Baker.)

Organizations

Many organizations other than the federal government are involved in the management of nuclear wastes and the provision of information and opinion. This section lists categories of them and gives some examples of each. For more comprehensive information, see the Buyers Guide issued annually by *Nuclear News*. Some 150 organizations in the United States and abroad are listed.

Commercial firms serve as contractors for national laboratories, e.g., Battelle for Pacific Northwest, Bechtel for Idaho, and Lockheed Martin for Sandia. Universities manage others, e.g., University of California for Los Alamos, Lawrence Livermore, and Lawrence Berkeley; and University of Chicago for Argonne. Combinations of industry and academia round out the picture, e.g., Battelle-University of Tennessee for Oak Ridge, and Battelle-State University of New York at Stony Brook for Brookhaven. Links to all of the DOE labs are found on the Internet (see References).

Firms also provide management of projects and facilities, e.g., Westinghouse for Yucca Mountain and Savannah River Site. They provide services, e.g., Duratek for treatment and disposal of low-level wastes.

The nuclear power field is represented by the Nuclear Energy Institute, the Electric Power Research Institute, and the Institute for Nuclear Power Operations. Information, research, and evaluation are provided.

Professional societies publish books and magazines and sponsor conferences on wastes. For example, the American Nuclear Society issues *Radwaste Solutions, Nuclear News*, and other technical publications. The Health Physics Society publishes its journal. Other societies that maintain cognizance of radioactive wastes are American Society of Mechanical Engineers, American Physical Society, Materials Research Society, American Chemical Society, American Institute of Chemical Engineers, American Ceramic Society, Institute of Electrical and Electronics Engineers, and Institute of Nuclear Materials Management. Three periodicals that deal with waste are *Waste Management* (Pergamon Press), *Radioactive Waste Management and the Nuclear Fuel Cycle* (Harwood, London), and *Radioactive Waste Exchange*. Battelle Press specializes in technical books on radioactive waste.

Nuclear engineering programs in universities provide one or more courses dealing with radioactive wastes. The catalog of schools with faculty contact information is maintained on the Web (see References).

The annual symposium on waste management held in Tucson, Arizona, publishes proceedings on recent problems and progress. Other important organizations are the American National Standards Institute, the National Academy of Sciences, the National Governors' Association, the National Conference of State Legislatures, and the Low-Level Radioactive Waste Forum. Worldwide waste management is addressed by the International Atomic Energy Agency in Vienna and the Organisation for Economic Co-operation and Development, headquartered in Paris. The Nuclear Energy Agency of the Organisation for Economic Co-operation and Development (OECD), headquartered in Paris, supplies many international publications on radiation management through its Web site (see References).

A number of public interest and environmental groups express concerns about radioactive wastes, among them the Union of Concerned Scientists, the National Resources Defense Council, Nuclear Energy Information Service, Nuclear Information and Resource Service, and Friends of the Earth. A comprehensive set of links to such organizations is found on the Web site of RadWaste.org (see References).

Investigations of Radiation Effects

Two projects have been initiated to investigate the exposure of members of the public to radiation. One has to do with radiological experiments on human beings; the other with fallout from weapons tests.

In 1993 it was revealed that, between 1945 and 1971, government agencies had conducted a number of radiological tests on human beings. Examples cited in the news media were: (a) releases of radioactivity to public areas as part of Cold War investigations of fallout and radiological weapons; (b) irradiation of prisoners with x-rays to determine doses that produce sterility; (c) injection of plutonium into terminally ill patients; and (d) use of radioactive tracers in food given to mentally retarded children for metabolism studies. The data from the research has been in the literature and has been used to determine safe and harmful levels of radiation. Those tests involving military applications were carried out in secret. Many of the studies were conducted without the knowledge or approval of the subjects, and some involved exposures now considered dangerous.

In 1994 President Clinton created the Advisory Committee on Human Radiation Experiments, with a charge to investigate the problem. The committee issued in 1996 a comprehensive report, available on the Web, along with certain data (see References). The report admits that many of the experiments were improper and recommends restitution to victims.

The United States and several other nations conducted aboveground weapons tests in the period 1951-1962, resulting in considerable fallout of radioactivity. The extent of harm to members of the public was first revealed in a 1997 report by the National Cancer Institute (see References). An extended study was initiated in 1998, and a very large preliminary report issued for review (see References). In summary of the findings, the number of cases of thyroid cancer as a result of iodine-131 and the number of premature deaths are larger than expected.

CHAPTER 23

Societal Aspects of Radioactive Wastes

We have seen how physical sciences and life sciences are involved in the management of radioactive wastes. Technical areas applied to ensure safe treatment and disposal are physics, chemistry, mathematics, engineering, geology, hydrology, seismology, climatology, meteorology, biology, ecology, and medicine. Despite the knowledge and talent available in these subjects, it has been difficult over the years to arrive at firm decisions and to achieve positive action in waste disposal. Part of the problem stems from the greater attention paid to the products of nuclear processes such as weapons and fuel for power plants rather than the by-products of wastes. Another part has been the many changes in direction through different national administrations. But a significant part of the problem has been public opposition.

This chapter describes the situation, noting some of the views and attitudes. Opinions vary with individual background, knowledge, personal experiences, and professional affiliation. In a democratic society such differences are to be expected. This chapter begins to examine the relationship of technical and societal aspects and to indicate possible means by which they may be brought into harmony.

Opposition to change has increased in recent years, along with a worldwide decline in acceptance of the authority of government and industry, and a demand for greater public involvement in decisions. Reactions are not limited to waste disposal by any means. Almost any new facility is viewed as a threat, whether it is a shopping center, low-income housing, a home for physically or mentally ill, a shelter for homeless people, a recycling plant, a waste incinerator, or a landfill. Perhaps the comments of Wildavsky (see References) are not too far off the mark: "How extraordinary! The richest, longest-lived, best-protected, most resourceful civilization, with the highest degree of insight into its own technology, is on its way to becoming the most frightened. Has there ever been, one wonders, a society that produced more uncertainty more often about everyday life? It isn't much really, in dispute—only the land we live on, the water we drink, the air we breathe, the food we eat, and the energy that supports us. Chicken Little is alive and well in America."

"NIMBY"

An almost universal reaction to a waste disposal facility (shown in the conceptual drawing) is, "Not in my back yard." This attitude is so prevalent that it is given the acronym "NIMBY." Several reasons give rise to the response, varying in intensity and ranging from the highly emotional to the very intellectual.

Artist's conception of a low-level radioactive waste disposal facility for the state of North Carolina. Concrete vaults are covered with earthen mounds planted with vegetation. (Courtesy of Chem-Nuclear Systems.)

First is fear of the unknown among those with no scientific knowledge. Radiation is regarded as mysterious and hence dangerous. Even among informed people, there is fear of anything nuclear. A large fraction of the people living today have grown up under the threat of nuclear warfare, and they tend to associate bombs, nuclear reactors, radioactivity, radiation, and wastes. The potential for physical damage by nuclear weapons and the ability of radiation to produce cancer are well known and dominate many people's thinking, even in situations where the hazard is minimal.

Second is concern that a waste disposal facility will fail and cause long-term contamination of air and water, with consequent hazard to people and the environment. Causes cited are human error in design or operation, or merely the validity of "Murphy's Law," which says that if anything can go wrong, it will.* The memory of the Three Mile Island and Chernobyl accidents is refreshed by those concerned about nuclear power. Some people may recognize that the potential danger is small but are concerned that they have no control over it, as they believe they have when driving an automobile, for example.

---

*Also see William and Mary Morris, *Morris Dictionary of Word and Phrase Origins*, Harper and Row, New York, 1972.

Third is the anticipation that the advent of a waste disposal facility will cause local land values to fall, will discourage new industry or tourism, or will change existing life styles. Such opinion can be modified by full understanding of benefits available to the region, including new tax and fee revenues and job opportunities.

Fourth is the opinion that waste disposal should be the responsibility of the power companies, as they are the principal generators. Believing that it is impossible to ensure isolation of wastes underground, many urge that wastes be stored above ground at nuclear stations. Long-term storage is said to allow new disposal options resulting from future research and development.

Fifth is distaste for being "imposed on" by having to accept a nuclear "dump," especially if those affected have had no voice in the decision. This reaction may be an extension of the frustration of observing local governments make changes in zoning, acquiescing to pressure from industry and developers. Government agencies have been perceived as insensitive to public concerns.

The sixth is a philosophic position, related to the belief that nuclear power is symbolic of centralized authority. Some opponents hold that the waste problem would disappear if all nuclear plants were shut down or phased out rapidly. A corollary is that the power supply should be replaced by stricter conservation measures and the use of renewable power sources involving sunlight and wood, with authority widely distributed in the hands of the public.

The response by the public to a waste disposal project takes various forms: letters to the administrator of the program, to editors of newspapers, or to government officials; speeches against the project at public meetings; voting elected representatives out of office; holding protest rallies; initiating injunctions and lawsuits; and making threats of physical violence. It is never clear, however, how widespread the resistance actually is, for two reasons: (a) the intensity of opposing rhetoric can be very misleading as to what is the real public opinion; (b) many people who support the endeavor or are neutral are afraid to express themselves.

Countering all of the above negative aspects is the realization by the public that nuclear medicine using radioactive materials and radiation saves countless lives, that nuclear power contributes one-fifth of our badly needed electrical power, and that nuclear reactors do not emit gases that pollute the atmosphere and contribute to potential climate change.

Public Participation

It is often noted that the solution to the nuclear waste problem is more social than technological. The narrowest interpretation of that statement is that it is necessary only to educate the public. Administrators who proceed solely on that assumption are doomed to failure.

Four levels of interaction are needed among those who undertake any project and those affected. Providing people with information is necessary, but not sufficient. Next is actively seeking public comment. Still higher in importance is giving the public opportunity to influence decision through an agency policy that takes heed of people's opinions and of the additional information they supply. The final step would be to submit agency proposals to the public for approval. This procedure would have to be limited to major actions, not day-to-day operations. Some observers maintain that public approval is the only way to success, while others believe that the establishment of confidence and trust in the project leaders is the key. The use of prestigious academic bodies to review and comment on issues and plans has been employed with some favorable results. A supervising body composed of citizens is probably preferable to a government agency, except for the difficulty in finding knowledgeable people to serve who are not in some way associated with nuclear activities. If technical individuals are responsible for a waste disposal project, they must be convinced of the need for public understanding and public participation in establishing disposal facilities. They must be able and willing to devote considerable time and effort to communication and consultation with environmental and public-interest groups, advisory committees, and individual citizens.

Scientists and engineers involved in waste management find themselves in awkward positions when they encounter public opposition. When asked to certify safety, they must answer in terms of probability, which does little to relieve doubts. When attacked verbally as being incompetent or being unconcerned about human values, some may be resentful and defensive, resulting in further alienation of the public. Technically knowledgeable people who are most successful in working with the public are those who seek to understand the worries and fears of citizens. At the same time, scientists and engineers, who have been trained to speak the truth as best they know it, must be aware that some zealots are willing to say whatever is likely to further their cause. They also should realize that government agencies and industrial organizations tend to release no more information than forced to, or to release favorable information, to "maintain a low profile" against attack. These practices tend to cause the public to mistrust government and industry. To maintain credibility and enhance trust, technical people in waste management must always provide full, accurate, and reliable information.

Facts and Ideas

Information that can be provided the public to alleviate unwarranted concerns includes the following:

1. Wastes are an inevitable by-product of our society's technological basis, including uses of the nucleus for medical diagnosis and treatment, electric power, and defense.

2. Much research and development has been directed towards safety measures in waste treatment, container design, transportation, and geologic disposal.

3. Strict regulations are in place for handling radioactive materials and for designing and operating high-level waste repositories and low-level waste disposal sites.

4. Radiation exposure to the public from transportation of wastes and from any disposal facilities is a small fraction compared to that from natural radiation.

5. A suitable choice of geologic medium has the capability of filtering radioactive elements and preventing contamination of water.

6. Radioactive materials eventually decay, in contrast with many permanent hazardous chemicals.

7. The longer the half-life, the smaller is the amount of radiation released from radioactive substances, all other things being equal.

8. Radioactive materials are readily detected by their radiation, in contrast with difficulty in measuring traces of chemicals.

9. More is known about the biological effects of radiation than of any other hazard.

10. Continuing to store radioactive wastes rather than disposing of them would leave a problem for future generations.

Mutual Understanding

It is possible to visualize an ideal situation, in which radioactive wastes are managed to the satisfaction of both the responsible agency and the public. Appropriate philosophies, policies, and practices to help achieve success can be identified.

A fact to be accepted is that radioactive wastes exist and will continue to be generated for some time, whether or not any new nuclear plants are built.

The ultimate goal is the safe management of wastes, to minimize hazard to workers and the public. Zero risk cannot be guaranteed, but the risk is small compared with those normally accepted as part of human life.

Even under conditions of negligible risk to people and the environment, there can be some adverse effects of a waste facility, requiring that some financial incentives or other benefits to a community be provided.

The agency charged with the task of choosing a disposal site or of managing a facility will make every effort to provide accurate technical information about the project to all interested parties.

It is understood by all that participation by the public in projects that affect human beings and the environment has become an expected supplement to the classic representative government. Discussion and decisions on actions on waste disposal are to be carried out openly, with opportunity for anyone to attend meetings and contribute opinions, and to have access to working documents.

Interaction between citizens and agency representatives is to be characterized by mutual courtesy and respect in an atmosphere that seeks to avoid antagonism. Those having legal responsibility for actions involving waste disposal will be responsive to comments, questions, suggestions, and recommendations by individual members of the public or by organized groups, but will make the ultimate decisions on steps to be taken. Those charged with the task of managing radioactive wastes will not be content with being mere administrators but will also strive to be leaders who instill confidence through actions rather than words.

There is historic precedent for the relationship of people and leaders. Recall the pronouncement of the Declaration of Independence, "governments are instituted among men, deriving their just powers from the consent of the governed."

CHAPTER 24

Perspectives—A Summing Up

This book has sought to provide a brief but broad picture of radioactivity, radiation, nuclear power, radiation applications, and radioactive waste management. It has tried to be informative, factual, and fair. To sum up, here are some highlights, key ideas, and important conclusions about nuclear wastes.

1. Radioactivity is both natural and artificial. The decay process gives radiation in the form of alpha particles, beta particles, and gamma rays. All of us are exposed to natural background radiation coming from cosmic rays and minerals in the ground. Radon is an important source of exposure.

2. Radiation can be harmful to the body and to genes, but an effect at low levels is questionable. Several methods of protection are available, including time, distance, and shielding.

3. The fission process yields energy that is converted into electricity in nuclear plants, but also generates wastes in the form of radioactive fission products and activation products. It is difficult and expensive to destroy wastes.

4. There is a large volume of defense wastes, stemming from World War II and subsequent weapons production. Its processing and disposal is underway by the Department of Energy. Weapons are dismantled and the material disposed as waste or used as nuclear reactor fuel.

5. Radioisotopes and radiation are used extensively in research, medical diagnosis and treatment, industry, and space applications. Many of the waste by-products are of short half-life.

6. Low-level radioactive waste comes from nuclear reactors, industry, and health-related institutions. Industry has greatly reduced waste volume over the years, but the number of waste burial sites available has dwindled. New disposal sites will be needed in the future.

7. The most radioactive of the wastes to be disposed of is spent fuel from nuclear power reactors. It continues to accumulate at reactors in water pools or dry storage containers.

8. Attempts have been made for many years to find suitable locations for the disposal of high-level wastes, including used nuclear fuel. Emphasis on power production and frequent changes in program direction have delayed high-level waste disposal.

9. The public is concerned about the transport of radioactive wastes, but isotopes, spent fuel, and other wastes are routinely transported in specially designed containers able to withstand conceivable accidents.

10. Plans for disposal set by Congress are to be carried out by the Department of Energy for spent fuel, and by the states for low-level waste. Decisions must be made about what is best to do with plutonium and highly enriched uranium produced by the United States and the former Soviet Union as part of the Cold War.

11. Any waste isolation system must protect the public and future generations from harm caused by radiation exposure. Design goals are to isolate reactor wastes and limit public exposure to a small fraction of normal background. Isolation need not be forever, but only until enough radioactive decay has taken place.

12. Waste management involves application of science and engineering, but also requires that the public participate in selecting disposal sites and technology.

13. Several methods for disposing of spent fuel have been studied, with the conclusion that geologic disposal in a mined cavity deep in the earth is preferable. Disposal in the seabed is a backup approach but unlikely to be used.

14. Interstate compacts were formed in the United States to handle all aspects of low-level waste disposal—site selection, design, construction, operation, site closure, and long-term care.

15. An underground repository for transuranic waste is in operation in New Mexico and progress is being made toward a repository for high-level waste at Yucca Mountain in Nevada.

16. Protection of the public and the environment from radiation exposure by radioactive wastes is provided by multiple barriers that include waste form, containers, and the geologic medium.

17. For low-level wastes, several alternatives to shallow land burial are intermediate-depth disposal, mined-cavity disposal, below-ground vault, modular concrete canister disposal, earth-mounded concrete bunkers, shaft disposal, aboveground vault, and earth-mounded aboveground vault. These secure designs are available but compacts have been unable to establish new disposal facilities.

18. A legal structure exists to manage wastes safely, including laws specifying policy, requirements, and schedule; environmental protection standards; and licensing and regulation by the federal government and the states.

19. Laboratory and field experiments and tests yield data for use in calculation of predicted performance of disposal systems, but they cannot prove that there will be zero release of radioactivity. Technical experts believe that the hazard can be kept far below that acceptable from other sources of risk in daily living.

20. Nuclear processes have become a significant part of our economy and our culture through defense use, medical applications, and the generation of electric power. Long-term energy demands will be met only by use of all available sources, including nuclear. Wastes are an inevitable by-product and must be safely isolated.

21. Continued study of the phenomena of global warming, acid rain, and ozone depletion will tell what environmental benefit derives from the use of nuclear power as a source of electricity.

22. It is a challenge to all of us to enhance the benefits and reduce the risks in the applications of the nucleus. This will involve dedicated and sincere efforts by program leaders to provide accurate and complete information, to understand and respond to concerns, and to establish meaningful dialog with those people who are affected.

Glossary

Brief and simple definitions are given for words and phrases in this book and other literature on waste management. This glossary is divided into four topics:

- Nuclear processes and radiation
- Reactors and fuel
- Waste characteristics
- Geologic features

Nuclear Processes and Radiation

Accelerator—an electric/magnetic device to give charged particles a very high speed.

Activation—producing radionuclides by irradiation, especially by neutrons.

Activity—rate of disintegration (also see **Curie**).

Alpha particle—a type of radiation; the helium nucleus.

Atomic number—number of electrons or protons in an atom.

Attenuation—reduction in radiation on passage through matter.

Becquerel (Bq)—unit of activity, 1 disintegration a second (d/s).

Beta particle—a type of radiation; the electron.

Biological half-life—time for half of a radioisotope to be eliminated from the body.

Curie (Ci)—a unit of radioactivity equal to 37 billion disintegrations a second (d/s).

Dating—finding the age of an object by its radioactivity.

Daughter—a nucleus that results from radioactive decay.

Decay—the disintegration process of nuclei.

Dose—amount of energy absorbed.

Electron—basic electrically charged particle, mass 9.1×10^{-31} kilograms, charge 1.6×10^{19} coulombs.

Electron volt (eV)—a unit of energy, 1.6×10^{-19} watt-seconds.

Fission—splitting of nuclei by neutrons.

Fission products—the nuclei, usually radioactive, resulting from fission.

Gamma ray—a type of radiation; a high energy photon or electromagnetic wave.

Genetic—an effect (as of radiation) on hereditary tissue.

Gray (Gy)—100 rads.

Half-life—the length of time for half the atoms of a radioactive substance to decay.

Hormesis—beneficial effect of small doses of radiation.

Ionization—removal of electrons from an atom, for example, by means of radiation.

Isotope—atoms with the same atomic number but different mass number.

Joule—a unit of energy, the watt-second.

Linear hypothesis—the assumption that any radiation causes biological damage, according to a straight-line graph of health effect versus dose.

Mass number—number of neutrons plus protons in a nucleus.

Neutron—a basic particle that is electrically neutral, weighing nearly the same as the hydrogen atom.

Periodic table—a chart of the chemical elements.

Person-rem (or man-rem)—the product average dose and the number of people affected.

Rad—a unit of radiation energy absorption; 1/100 joule per kilogram.

Radiation—particles or waves from atomic or nuclear processes (or from certain machines).

Radioactivity—spontaneous disintegration of an unstable nucleus.

Radioisotope—a radioactive isotope.

Radionuclide—a species of atom that is radioactive.

Rem—unit of radiation dosage equal to the rad for x-rays, gamma rays, and some beta particles; accounts for biological effect.

Sievert (Sv)—100 rems.

Somatic—a direct effect (as of radiation) on the health of tissue.

Tracer—an isotope used to follow a process.

X-rays—electromagnetic radiation of energy greater than that of visible light, usually produced by an x-ray machine.

Reactors and Fuel

Assembly—bundle of fuel rods used in a reactor.

Barrier—metal sieve used in gaseous diffusion isotope separation.

Boiling water reactor (BWR)—a light-water cooled reactor in which some boiling occurs.

Breeder reactor—a reactor that produces more fissile material than it consumes (by a process called "breeding").

Cladding—the outer coating of nuclear fuel, for example, a tube.

Converter—a reactor in which some fertile material is made into fissile material.

Criticality—a condition in which a chain reaction involving neutrons and fuel is self-sustaining.

Decommissioning—removal from service at the end of useful life.

Decontamination—the removal of radioactive material.

Enrichment—a process to increase the percentage of a desired isotope, such as uranium-235.

Fabrication (of fuel assemblies)—making uranium oxide fuel pellets and forming fuel rods and bundles of rods.

Fertile—a material that becomes fissile upon absorbing a neutron.

Fissile—able to be split by a low-energy neutron.

Fuel—fissionable material "burned" in a nuclear reactor, for example, uranium.

Fuel cycle—all steps in supplying, using, and processing fuel for nuclear reactors, including disposal of wastes.

Fusion—a nuclear process in which nuclei are combined to yield energy.

Implosion—a compression to detonate a nuclear weapon.

Licensing—giving a permit to build or operate a facility.

Light-water reactor (LWR)—a nuclear reactor cooled and moderated by water.

Metric ton—see **Tonne**.

Mixed oxide—mixture of oxides of uranium and plutonium.

Moderator—a light element used to slow neutrons.

Multiplication—neutron interaction with fissile material in a chain reaction.

Natural uranium—uranium as mined (0.7% ^{235}U, 99.3% ^{238}U).

Pile—an early name for nuclear reactors.

Plutonium—the element formed by neutron absorption in uranium-238.

Pressurized water reactor (PWR)—a light-water cooled reactor operated at high pressure without boiling.

Reactor—a device involving a chain reaction using neutrons.

Spent fuel—nuclear fuel that has been removed from a reactor after use to produce power.

Tonne—a metric ton, 1000 kilograms.

Waste Characteristics

Actinides—elements of the periodic table with atomic numbers 90 through 103.

Barrier—a component that slows the movement of radioisotopes.

Calcine—powder produced by heat treatment.

Canister—the primary container for solid waste.

Cask—container for fuel assemblies.

Ceramic—insoluble solid oxide.

Disposal—removal from the environment permanently.

Grout—a cement mixed with wastes.

Ion exchange—a process used to purify chemicals.

Isolation—preventing migration of wastes to the biosphere.

Leaching—dissolving in a liquid.

Low-level wastes—those generally not requiring shielding or heat removal; small transuranic content.

Mill tailings—see **Tailings**.

Once-through—a fuel cycle in which spent fuel is not reprocessed.

Partitioning—separation of certain radioisotopes from waste.

Regulation—maintenance of standards of performance through rules.

Repository—a location for waste to be disposed of.

Reprocessing—the mechanical and chemical treatment of nuclear fuel to separate uranium, plutonium, and fission products.

Retrievable—able to reclaim if necessary.

Salt—sodium chloride, NaCl, as a geologic medium.

Storage—holding temporarily.

Tailings—the residue from extraction of uranium from its ore.

Transmutation—transformation of isotopes using nuclear reactions.

Transuranic—beyond uranium in the periodic table.

TRU—transuranic waste, with more than 10 nanocuries a gram.

Vitrification—solidification of wastes with glass.

Geologic Features

Aquifer—underground layer of material through which water passes.

Bed—layered deposit of sediment in the form of rocks, products of weathering, organic materials, and precipitates. Also, **bedded**.

Biosphere—regions of the earth and atmosphere occupied by living beings.

Breccia—fragmented rock region (as in Breccia pipe).

Dome—a bed that arches up to form a rounded peak deposit. Also, **domed**.

Drift—waste emplacement opening.

Fault—a break in a rock formation usually involving diagonal movement. An example: the San Andreas fault in California.

Fold—a curved deformation of rock. The peaks are called anticlines; the valleys, synclines.

Glacier—large body of ice, often moving slowly.

Igneous rocks—formed by solidification of molten rock.

Lava—molten rock that issues from a volcano.

Leaching—dissolving.

Magma—molten rock within the earth.

Metamorphic rocks—those changed by temperature and pressure.

Meteorite—a solid body from outer space that reaches the earth without vaporizing.

Percolation—movement of water through openings.

Permeability—capacity of a rock, sediment, or soil for transmitting fluids.

Porosity—the ratio of the total volume of small spaces or pores in a rock or soil to its total volume.

Salt bed—a deposit formed by the evaporation of sea water.

Sedimentary rocks—deposited in layers near the surface by water, wind, and ice.

Seepage—slow movement of water.

Tectonic plate—geological concept of the movement of large segments of the earth's crust.

Total system performance assessment—demonstration of assurance of safe isolation of wastes.

Unsaturated zone—underground layer containing air and water.

Viability assessment—basis for study of a repository site.

Volcano—a vent in the crust of the earth from which lava, gases, and ash erupt.

Water table—top of an aquifer.

Reference Material on Wastes

There is an enormous body of literature on radioactive wastes, much of it originating in government agencies. Presented here first are "General References," including sources of scientific information and comprehensive works on nuclear energy and radioactive wastes.

The Internet has become the principal source of information for many people. There are several reasons: a library may not be nearby; if there is a library, it might not stock the needed reference; if it has the reference, the copy may be checked out. In contrast, search engines such as Google are at one's fingertips, yielding useful World Wide Web site links instantly. Disadvantages are the lack of peer review and the tendency for sites to vanish. In this edition of the book, we have sought to cite relevant and robust Web sites for supplemental reading. In the chapters, for convenience, references to Web sites are separate from those of print media.

General

Nuclear Science and Technology and How It Influences Your Life
  http://www.aboutnuclear.com
  Public information by the American Nuclear Society.
American Nuclear Society publications
  http://www.ans.org Nuclear News, Radwaste Solutions, Nuclear Technology, Nuclear Science and Engineering, Fusion Science & Technology, and Transactions of the American Nuclear Society.
Waste Management Symposium.
  Held annually in Tucson, AZ.
  http://www.wmsym.org/Home page.
WWW Virtual Library
  http://www.vlib.org
  Links to virtual libraries in engineering, science, and other categories.
Nuclear Engineering, University of California at Berkeley
  http://www.nuc.berkeley.edu
  Select Links/Virtual Library: Nuclear Engineering.
Radiation Information Network
  http://www.physics.isu.edu/radinf/index1.html
  Numerous links to sources. By Idaho State University Health Physics Program.
Scientific American articles
  http://www.scientificamerican.com
  Select Archive for articles from 1993 to present. Use "Full Text Search" feature with topic Radioactive Waste.
How Things Work
  http://howthingswork.virginia.edu

Information on many subjects by Professor Louis Bloomfield.
How Stuff Works
http://www.howstuffworks.com
Brief explanations by Marshall Brain of familiar devices and concepts, including many of the topics of this book.
Scout Report
http://scout.cs.wisc.edu
Provides links to many subjects.
Energy Quest
http://www.energyquest.ca.gov
An energy education Web site from California Energy Commission.
Access to government science programs
http://www.science.gov
Example: Select Environment & Environmental Quality/wastes (solid or concentrated)/Radioactive waste.
Glossary of nuclear terms
http://www.nrc.gov/reading-rm/basic-ref/glossary.html
Informative definitions of terms from Nuclear Regulatory Commission. Includes references to other glossaries.

Glossary of Terms in Nuclear Science and Technology, American Nuclear Society, La Grange Park, IL, 1986.

The League of Women Voters Education Fund, *Nuclear Waste Primer*, Lyons & Burford, New York, 1993. Main chapters: Basic Information, Civilian High-Level Waste, Civilian Low-Level Waste, Transportation and Liability, Defense Waste, and Roles for Citizens. Easy reading but slightly out of date.

Raymond L. Murray, *Nuclear Energy*, 5th Ed., Butterworth-Heinemann, Boston, 2001. Subtitle: An Introduction to the Concepts, Systems, and Applications of Nuclear Processes. A description of basic nuclear phenomena, devices, and processes, followed by a discussion of problems and opportunities. Includes applications of isotopes and radiation, nuclear reactor operation, and waste disposal. Designed for reading by college students or technically oriented laypeople.

David Bodansky, *Nuclear Energy: Principles, Practices, and Prospects*, American Institute of Physics, Woodbury, NY, 1996. All aspects of the subject from a generally non-mathematical view.

Albert B. Reynolds, *Bluebells and Nuclear Energy*, Cogito Books, Madison, WI, 1996. A readable description of radiation, fission, reactors, and wastes. For teachers, students, and the public (bluebells are incidental).

Harry Henderson, *Nuclear Power: A Reference Handbook*, ABC-CLIO, Santa Barbara, CA, 2000. A compendium of information leaning toward the antinuclear. Nuclear chronology, biographical sketches, U.S. nuclear plants, organizations, and references (print, Web, and video).

Ronald Allen Knief, *Nuclear Engineering: Theory and Technology of Commercial Nuclear Power*, 2nd Ed., Taylor & Francis, London, 1992. Thorough coverage at an advanced level of nuclear reactor systems, reactor safety, the nuclear fuel cycle, and nuclear fusion.

Robert E. Berlin and Catherine C. Stanton, *Radioactive Waste Management*, John Wiley & Sons, New York, 1989. Describes radioactive waste forms, regulations, processing of low-level wastes, disposal of all kinds of waste, and remediation of contaminated sites.

Foo-Sun Lau, *Radioactivity and Nuclear Waste Disposal*, John Wiley & Sons, New York, 1987. Section 1 contains fundamentals; Section 2 covers waste management, including programs around the world.

James H. Saling and Audeen W. Fentiman, *Radioactive Waste Management*, 2nd Ed., Taylor & Francis, New York, 2001. Treats each waste type separately—spent fuel, HLW, TRU, LLW, and mill tailings. Also covers transportation and decontamination/decommissioning. Contains many tables, data, diagrams, graphs, and references. A recommended acquisition.

Radioactive Waste Management: An IAEA Source Book, International Atomic Energy Agency, Vienna, 1992. Designed for reading by the public, this book covers technical and institutional aspects, sociopolitical and ethical concerns, public understanding and acceptance, and international organizations.

A. Alan Moghissi, Herschel W. Godbee, and Sue A. Hobart, *Radioactive Waste Technology*, American Society of Mechanical Engineers, New York, 1986. Sponsored by ASME and the American Nuclear Society. Useful information on all aspects of wastes.

Isaac Asimov, *Asimov's Biographical Encyclopedia of Science and Technology*, 2nd Ed., Doubleday, New York, 1982. Reprint by New Library Press, 2001. Subtitle: The Lives and Achievements of 1510 Great Scientists from Ancient Times to the Present Chronologically Arranged.

Grolier Multimedia Encyclopedia, Grolier, Inc., Danbury, CT. Annual update. CD-ROM that includes articles on nuclear energy and fusion energy.

Chapter 1—Questions and Concerns About Wastes

United Nations Framework Convention on Climate Change and the Kyoto Protocol
http://www.unfccc.int
Includes full texts of the documents.

United Nations Summit 2002
http://www.johannesburgsummit.org
Official Web site on sustainable development. Link to follow-up.

Global Warming
http://www.uic.com.au/nip24.htm
A briefing paper from Australia.

Congressional Research Service (National Council for Science and the Environment)
http://www.cnie.org
There is a CRS on which one must click.

Basic Physics Course
http://zebu.uoregon.edu/~kevan/found.html
A complete course given by Professor Stephen Kevan. Try the lectures of Nov. 1, 1995 (Newton's laws), and Oct. 11, 1995 (special relativity).

PhysLink
http://www.physlink.com
Select Reference for links to sources of many physics constants, conversion factors, and other data.

General Chemistry
http://www.chem.ufl.edu/~chm2040/Notes
An online course from the University of Florida.

Charles H. Heimler and Jack Prince, *Focus on Physical Science*, Merrill Publishing Co., Columbus, OH, 1989. One of several secondary school texts that contain interesting background information. This book emphasizes investigations by the student.

Peter Alexander, et al., *Physical Science*, Silver, Burdette, and Ginn, Morristown, NJ, 1990. Another basic text with attractive photographs and diagrams, simple mathematics.

David R. Lide, Ed., *CRC Handbook of Chemistry and Physics*, CRC Press, Boca Raton, FL. Standard reference for physical data and mathematical information. Issued annually.

Henry N. Wagner, Jr., and Linda E. Ketchum, *Living with Radiation: The Risk, the Promise*, Johns Hopkins Press, Baltimore, 1989. Reprint by Books on Demand, 1999. Thoughtful and informative essays on radioactivity, radiation, weapons, nuclear medicine, and nuclear reactor safety.

Chapter 2—Atoms and Chemistry

Educational Site
http://www.lbl.gov/Education/index.html
Select ABC's of Nuclear Science; California Science Project; Glenn Seaborg: His
Life & Contributions.
The Particle Adventure
http://www.cpepweb.org
A leisurely and light-hearted tour of quarks, antimatter, the Standard Model,
interactions, and experiments.
WebElements Periodic Table of the Elements
http://www.webelements.com
Mark Winter provides information about each element.
Frederick E. Trinklein, *Modern Physics*, Holt, Rinehart, and Winston, Austin, TX,
1990. A high school text in basic physics with good chapters on atomic structure,
nuclear reactors, and high-energy physics.
Paul Allen Tipler and Ralph A. Llewellyn, *Elementary Modern Physics*, 4th Ed., W.
H. Freeman, San Francisco, 2002. A college text in atomic and nuclear physics for
students with a background in basic physics.
John R. Taylor and Chris D. Zafiratos, *Modern Physics for Scientists and
Engineers*, Prentice-Hall, Englewood Cliffs, NJ, 1991. Another excellent text in
college atomic and nuclear physics.
Samuel Glasstone and Alexander Sesonske, *Nuclear Reactor Engineering*, 4th Ed.,
Chapman and Hall, New York, 1994. A classic reference on the subject.

Chapter 3—Radioactivity

ABC's of Nuclear Science
http://www.lbl.gov/abc
Radioactivity, radiations, and suggested experiments.
Radioactivity in Nature
http://www.physics.isu.edu/radinf/natural.htm
Facts and data on natural and manmade radioactivity and on radiation. From
Idaho State University Physics Department.
Radioactive Decay Series
http://www.uic.com.au/neAp2.htm
Diagrams of uranium, thorium, and actinium series. Part of an online book
Nuclear Electricity, accessible from this site. By Uranium Information Centre,
Australia.
History of Radioactivity
http://www.accessexcellence.com/AE/AEC/CC
Biographies of scientists responsible for discoveries, by Genentech.
Environmental Radiation and Uranium: A Radioactive Clock
http://scifun.chem.wisc.edu/chemweek/Radiation/Radiation.html
Briefings on radiation and its effect; uranium decay series.
W. B. Mann, A. Rytz, A. Spernol, and W.L. McLaughlin, *Radioactivity and Its
Measurement*, 3rd Ed., Butterworth-Heinemann, Boston, 1991.
Handbook of Radioactivity Measurement Procedures, NCRP Report No. 58, 2nd
Ed., National Council of Radiation Protection and Measurements, Bethesda, MD,
1985.
Egardo Browne and Richard Firestone; Virginia Shirley, Ed., *Table of Radioactive
Isotopes*, John Wiley & Sons, New York, 1986.
Norman Holden, "Table of the Isotopes," in *CRC Handbook of Chemistry and
Physics* (See Chapter 1 references). For each isotope, gives natural abundance,
atomic mass or weight, half-life, energies of radiation, and thermal neutron cross
sections. Updated periodically.
Robley D. Evans, *The Atomic Nucleus*, Krieger Publishers, Melbourne, FL, 1982. A
reprint of a classic 1955 advanced textbook.

Chapter 4—Kinds of Radiation

National Institute of Science and Technology (NIST)
http://www.nist.gov
For stopping powers and ranges of electrons, protons, and alpha particles, use Search with keyword "stopping power." For photon cross sections for many elements, use Search with keyword "XCOM." Especially look for data by Berger and Hubbell.

P. N. Cooper, *Introduction to Nuclear Radiation Detectors*, Cambridge University Press, Cambridge, England, 1986. A small book, easily read. No extensive knowledge of nuclear physics is assumed.

Glenn F. Knoll, *Radiation Detection and Measurement*, 3rd Ed., John Wiley & Sons, New York, 1999. A comprehensive and readable college textbook.

Merril Eisenbud and Thomas F. Gesell, *Environmental Radioactivity from Natural, Industrial, and Military Sources*, 4th Ed., Academic Press, San Diego, CA, 1997.

Chapter 5—Biological Effects of Radiation

Radiation and Health Physics
http://www.umich.edu/~radinfo
Many links. By University of Michigan Student Chapter of the Health Physics Society.

Radiation Dose and Biological Effects
http://www.hps.org
Select Public Information, Radiation Facts for pdf files. From Health Physics Society.

Health Effects of Low-Level Radiation
http://www.ans.org/pi/ps
Select Position Statement. By American Nuclear Society.

How Does Radiation Affect the Public
http://www.nrc.gov/what-we-do/radiation/affect.html
Information from Nuclear Regulatory Commission.

Low Dose Radiation Research Program of DOE
http://www.er.doe.gov/production/ober/lowdose.html
Description of program.

Biological Effects of Low Level Exposures (BELLE)
http://www.belleonline.com
Description of organization and links to literature.

"Validity of the Linear No-Threshold Theory of Radiation Carcinogenesis at Low Doses"
http://www.world-nuclear.org/sym/ 1998/cohen.htm
Bernard L. Cohen's paper at a 1998 symposium of the Uranium Institute.

Bernard L. Cohen, "Catalog of Risks Extended and Updated," *Health Physics 68*, 157, 1995. Also found in Wilson and Crouch article, pp. 211-212.

T. O. Tengs, et al., "Five Hundred Life Saving Interventions and Their Cost Effectiveness," *Risk Analysis 15*, 369, 1995. Also found in Wilson and Crouch article, pp. 244-263.

R. E. J. Mitchel, "Low-Dose Radiation Risk: A Biological Reality Check," *Radwaste Solutions*, March/April 2002. Describes experiments and suggests mechanisms for the responses to radiation that deviate from the linear hypothesis.

Jacob Shapiro, *Radiation Protection: A Guide for Scientists and Physicians*, 4th Ed., Harvard University Press, Cambridge, MA, 2002. Technical but not highly mathematical; comprehensive yet practical.

Eric J. Hall, *Radiation and Life*, 3rd Ed., McGraw-Hill, New York, 1993. Sources of radiation and biological effects.

Edward Pochin, *Nuclear Radiation: Risks and Benefits*, Oxford University Press, Oxford, England, 1992. Discusses cancer and gene damage.

Herman Cember, *Introduction to Health Physics*, 3rd Ed., McGraw-Hill Medical Division, New York, 1996.

James E. Turner, *Atoms, Radiation, and Radiation Protection*, 2nd Ed., John Wiley & Sons, New York, 1995. Text provides necessary scientific background as well as health physics information.

Chapter 6—Radiation Standards and Protection

Standards for Protection Against Radiation
http://www.nrc.gov/reading-rm/doc-collections/cfr/part020
Complete text of Part 20 of the Code of Federal Regulations, Title 10 Energy.
Bernard Schleien, Lester A. Slaback, Jr., and Brian Kent Birky, *Handbook of Health Physics and Radiological Health*, 3rd Ed., Lippincott Williams & Wilkins, Philadelphia, 1998. A collection of data on radiation protection, including the topics of doses, regulations, shielding, biology, and medicine. Useful information for health physicists.
BEIR Reports (from the Committee on Biological Effects of Ionizing Radiation). Sponsored by the National Research Council and published by the National Academy Press. Some can be read on line at http://www.nap. edu. Titles are:
  The Effects on Populations of Exposure to Low Levels of Ionizing Radiation, BEIR I (1972), BEIR II (1977), BEIR III (1980).
  Health Risks of Radon and Other Internally-Deposited Alpha-Emitters, BEIR IV (1988).
  Health Effects of Exposure to Low Levels of Ionizing Radiation, BEIR V (1990).
  Health Effects of Exposure to Radon, BEIR VI (1999).

Chapter 7—Fission and Fission Products

National Nuclear Data Center
http://www.nndc.bnl.gov
Maintained by Brookhaven National Laboratory. Links to many sources of data.
Basics of Nuclear Physics and Fission
http://www.ieer.org/reports/n-basics.html
Decay, binding energy, fission. By Institute for Energy and Environmental Research.
Donald C. Stewart, *Data for Radioactive Waste Management and Nuclear Applications*, John Wiley & Sons, New York, 1985. Major sections are Physical Data (fission product properties and chains, radioisotopes, and the ORIGEN computer program); Chemical Data; Radioactive Wastes; Data for Operations.
A. Michaudon, Ed., *Nuclear Fission and Neutron-Induced Fission Cross Sections*, Pergamon Press, New York, 1981. An advanced book on the fission process and methods of measurement.

Chapter 8—The Manhattan Project

The Smyth Report online
http://nuketesting.enviroweb.org/hew/Smyth
Preface, selected chapters, and appendices.
Federation of American Scientists
http://www.fas.org
Search on Manhattan Project for many links.
"The Italian Navigator Has Landed in the New World"
http://www.lanl.gov/worldview/welcome/history/08_chicago-reactor.html
Brief account of the first chain reaction.
Manhattan Project Heritage Preservation Association
http://www.childrenofthemanhattanproject.org
A valuable source of information.
F. G. Gosling, "The Manhattan Project: Making the Atomic Bomb"
http://www.osti.gov/bridge
Select Easy Search/Identifier Numbers; enter 303853. File 7311K in pdf.

History of Hanford Reactors
  http://www.hanford.gov
  Select History/The Hanford Site Historical District/Historic overview.
Department of Energy, 1977-1994: A Summary History
  http://www.osti.gov/bridge
  Select Easy Search/Identifier Numbers; enter DOE/HR-0098. Includes material
  from 1939 on. By Terrence R. Fehner and Jack M. Holl.
Nuclear Age Timeline
  http://www.em.doe.gov/timeline
  Events from the pre-1940s to the 1990s, with sources of additional information.
Corbin Allardice and Edward R. Trapnell, "The First Pile," *Nuclear News,* November
  2002, p. 34. An adaptation of a pamphlet written in 1946 recording the startup of
  the first reactor in 1942. Part of a 60th anniversary special section of the
  magazine.
Robert Jungk, *Brighter Than a Thousand Suns*, Harvest Books, Fort Washington,
  PA, 1970. A readable history of the period 1918-1955, including the development
  of the first nuclear weapons.
Stephane Groueff, *Manhattan Project*, Little, Brown & Co., Boston, 1967. Subtitle:
  The Untold Story of the Making of the Atomic Bomb. A highly-praised account of
  the period.
Henry D. Smyth, *Atomic Energy for Military Purposes*, Princeton University Press,
  Princeton, NJ, 1945. Reprint by Stanford University Press, 1990, with a new
  foreword by Philip Morrison and an essay by H. D. Smyth. The first public
  information about the Manhattan Project.
Richard Rhodes, *The Making of the Atomic Bomb*, Simon & Schuster, New York,
  1986. Thorough and well-written description of the World War II project.

Chapter 9—Defense and Development

U.S. Department of Energy laboratories
  http://phonebook.doe.gov/natlabs.html
  Links to home pages of all laboratories.
Nautilus Museum
  http://www.ussnautilus.org
  History, tour, and links.
U.S. Navy Fact File
  http://www.chinfo.navy.mil/navpalib/factfile/ffiletop.html
  Select Submarines or Surface Ships/ Aircraft Carriers.
Adams Atomic Insights
  http://ans.neep.wisc.edu/~ans/point_source/AEI/AEI_home.html
  Select Topical Index and articles on nuclear aircraft, nuclear ships (including *N.S.
  Savannah*), and nuclear batteries.
Bertrand Goldschmidt and Gerard Smith, *The Atomic Complex*, American Nuclear
  Society, La Grange Park, IL, 1982. Subtitle: A Worldwide Political History of
  Nuclear Energy.
Stelio Villani, *Isotope Separation*, 2nd Ed., American Nuclear Society, La Grange
  Park, IL, 1976. Describes most of the techniques.
Richard G. Hewlett and Francis Duncan, *Nuclear Navy 1946-1962*, University of
  Chicago Press, Chicago, 1974.
Frank G. Dawson, *Nuclear Power: Development and Management of a Technology*,
  University of Washington Press, Seattle, 1976. Covers the period 1945-1975.
Dietrich Schroeer, *Science, Technology, and the Nuclear Arms Race*, John Wiley &
  Sons, New York, 1984.
Controlled Nuclear Chain Reaction: The First Fifty Years, American Nuclear
  Society, LaGrange Park, IL, 1992. Reviews the startup of the first reactor at
  Chicago and discusses the development of nuclear power, peaceful applications,
  and nuclear safety.

Chapter 10—Uses of Isotopes and Radiation

Biology Links
    http://mcb.harvard.edu/BioLinks.html
    Harvard University Department of Molecular and Cellular Biology.
DOE Information Bridge
    http://www.osti.gov/bridge
    Search on A Vital Legacy PUB-800 for atoms in biology. File 4694K pdf.
Radioisotopes in Medicine
    http://www.uic.com.au/nip26.htm
    Uses in diagnosis and therapy. From Australia.
Smoke Detectors and Americium
    http://www.uic.com.au/nip35.htm
    Formation of americium-241 and its use.
Radiocarbon dating
    http://www.radiocarbondating.com
    Select Radiocarbon WEB-info for extensive information.
"An Overview of Neutron Activation Analysis"
    http://www.Missouri.edu/~glascock/naa_over.htm
    Article by Michael D. Glascock, Missouri University Research Reactor.
North Carolina State University's Nuclear Science Center
    http://www.ne.ncsu.edu/NRP/services.html
    Services for neutron activation analysis, neutron radiography, and radioisotope
    production.
Boron Neutron Capture Therapy
    http://www.virtualtrials.com/bnct.cfm
    Clinical trials and results. Article by Jaime G. Gomez, M.D.
Boron Neutron Capture Therapy
    http://www.physics.isu.edu/~rajat/bnct.html
    Accelerator program at Idaho State University.
Food irradiation by electrons
    http://www.extension.iastate.edu/foodsafety/rad/laf.html
    Diagram and data on linear accelerator.
Iowa State Food Safety Project
    http://www.extension.iastate.edu/foodsafety/rad/irradhome.html
    Information and links.
 Food Irradiation Information from Food Safety and Inspection Service
    http://www.fsis.usda.gov/oa/topics/irrmenu.htm
    Select Irradiation of Meat Food Products for access to USDA Final Rule in *Federal
    Register*, December 23, 1999.
Facts about Food Irradiation
    http://www.iaea.or.at/worldatom/Press/Booklets/foodirradiation.pdf
    From the International Atomic Energy Agency. File 1226K, pdf.
History of Screwworm Eradication
    http://www.nal.usda.gov/speccoll/screwworm/history.htm
    Discovery and application of Sterile Male Technique. From National Agricultural
    Library.
Eradication of tsetse fly in Zanzibar
    http://www.fao.org/NEWS/1998/980505-e.htm
    http://www.iaea.org/worldatom/Press/P_release/2002/prn0202.shtml
    News items from Food and Agriculture Organization of the United Nations and
    IAEA on a successful program.
Nuclear Power in Space (Two versions of DOE/NE-0071)
    http://www.nuc.umr.edu/nuclear_facts/spacepower/spacepower.html
    http://www.ne.doe.gov
    Select Public Information/Documents/report title.

Voyager Web site
  http://voyager.jpl.nasa.gov
  History and status of the spacecraft.
Radioisotope Power Systems
  http://www.ne.doe.gov/space/space-history.html
  History and description of RTG.
The *Galileo* Home Page
  http://www.jpl.nasa.gov/galileo
  Photos and information about the spacecraft.
National Homeland Security Knowledgebase
  http://www.twotigersonline.com/resources.html
  Links to a large number of sites related to security.
Calculating the New Global Nuclear Terrorism Threat
  http://www.iaea.org/worldatom/Press/P_release/2001/nt_pressrelease.shtml
  Comments by the Director General, IAEA.
Case Study: Accidental Leakage of Cesium-137 in Goiania, Brazil, in 1987
  http://www.nbc-med.org/SiteContent/MedRef/OnlineRef/CaseStudies/csgoiania.html
  Thorough review of the incident and its consequences. By Alex Neifert.
"Campaign Started to Eliminate Tsetse Fly" *Nuclear News*, April 2002, p. 59.
Gopal B. Saha, Jr., *Fundamentals of Nuclear Pharmacy*, 4th Ed., Springer-Verlag, New York, 1998. Textbook on radioactive substances used in nuclear medicine. Includes basics of radioactivity, production of radionuclides, detection of radiation, labeling of compounds, and information on uses of radiopharmaceuticals.
Michael F. L'Annunziata, *Radionuclide Tracers: Their Detection and Measurement*, Academic Press, London, 1987. Emphasizes the detection process rather than the applications. Many references are listed.
Ronald L. Eisenberg, *Clinical Imaging: An Atlas of Differential Diagnosis*, 4th Ed., Lippincott Williams & Wilkins, Hagerstown, MD, 2002. Contains a large number of illustrations of scans with various techniques.
John Harbert and Antonio F. G. da Rocha, *Textbook of Nuclear Medicine*, Vol. 1, Basic Science, Vol. 2, Clinical Applications, 2nd Ed., Lea and Febiger, Philadelphia, 1984. A relatively readable medical book.
Joseph A. Angelo, Jr. and David Buden, *Space Nuclear Power*, Krieger Publishing, Melbourne, FL, 1985. Includes isotopic power sources.
Making the Nation Safer: The Role of Science and Technology in Countering Terrorism, National Academy Press, Washington, D.C., 2002. Can be downloaded in pdf by chapters at http://www.nap.edu. Discusses the application of existing technologies, research opportunities, and organizational actions to protect U.S. infrastructures.
Management of Terrorist Events Involving Radioactive Material, National Council on Radiation Protection and Measurements, Bethesda, MD, 2001. The report was prepared before September 11, 2001. Considers weapons and radioactive materials. Reviews both medical and psychosocial effects.

Chapter 11—Classification of Wastes

The League of Women Voters Education Fund, *Nuclear Waste Primer*, The League of Women Voters Education Fund, Lyons & Burford, New York, 1993. An elementary and readable survey of the topic of radioactive wastes.
Edward L. Gershey, Robert C. Klein, Esmeralda Party, and Amy Wilkerson, *Low-Level Radioactive Waste: From Cradle to Grave*, Van Nostrand Reinhold, New York, 1990. Identifies all kinds of radioactive wastes and gives the history of commercial LLW disposal. Covers processing and disposal.

Chapter 12—Spent Fuel from Nuclear Reactors

Anthony V. Nero, Jr., *A Guidebook to Nuclear Reactors*, University of California Press, Berkeley, 1993. Excellent diagrams and photographs.

Frank J. Rahn, Achilles G. Adamantiades, John E. Kenton, and Chaim Braun, *A Guide to Nuclear Power Technology*, John Wiley & Sons, New York, 1984. Reprint by Krieger Publishing, Melbourne, FL, 1992. A very large amount of information in the form of text, tables, diagrams, and photographs.

Raymond G. Wymer and Benedict L. Vondra, Jr., Ed., *Light Water Reactor Nuclear Fuel Cycle*, CRC Press, Boca Raton, FL, 1981.

J. C. Wagner, *Computational Benchmark for Estimation of Reactivity Margin from Fission Products and Minor Actinides in PWR Burnup Credit*, Oak Ridge National Laboratory, Oak Ridge, TN, NUREG/CR-6747, 2001.

Chapter 13—Storage of Spent Fuel

Civilian Nuclear Spent Fuel Temporary Storage Options
http://cnie.org/NLE/CRSreports/Waste/waste-20.cfm
Article by Mark Holt of Congressional Research Service, 1998.

Private Fuel Storage, LLC
http://www.privatefuelstorage.com
Select About the Project/Transportation; read FAQs. Subtitle: Developing Safe, Clean, Temporary Storage for Spent Fuel From Commercial Nuclear Power Plants.

"Regulating Dry Cask Storage," *Radwaste Solutions*, July/August 2000, p. 10. Interview of Susan Shankman and Randy Hall by Nancy Zacha. Notes use of casks in more than a dozen sites, with more to be built.

Charles W. Pennington and Michael S. McGough, "MADness and Spent-Fuel Cask Safety," *Radwaste Solutions*, May/June 2002, p. 25. MADness is defined as a militant act of destructiveness. The paper describes the analysis of a collision between an aircraft and a dry storage cask.

Scott Northard, "Private Fuel Storage: Finding; Real Solutions for Centralized Spent-Fuel Storage," *Radwaste Solutions*, January/February 2000, p. 35. History, status, and future of the project.

A. M. Platt, J. V. Robinson, and O. F. Hill, *The Nuclear Fact Book*, 2nd Ed., Gordon and Breach, New York, 1985. A wealth of data on all aspects of energy, the nuclear fuel cycle, and nuclear wastes.

Chapter 14—Reprocessing, Recycling, and Resources

EBR-II
http://www.anlw.anl.gov/anlw_history/reactors/ebr_ii.html
History of reactor with links to safety tests.

Liquid Metal Fast Breeder
http://www.ans.neep.wisc.edu/~ans/point_source/AEI/oct95/LMFBR_oct95.html
History and status as viewed by Rod Adams in issue of *Atomic Energy Insights*.

Japanese prototype fast breeder reactor MONJU
http://www.jnc.go.jp/zmonju/mjweb/index.htm
Includes information on many aspects of FBRs around the world.

La Hague Reprocessing Plant
http://www.cogemalahague.fr
"We have nothing to hide from you." In French.

The Unofficial IFR Home Page
http://www.nuc.berkeley.edu/designs/ifr/
Select links for system diagrams and other information. By University of California.

Integral Fast Reactor
http://www.pbs.org/wgbh/pages/frontline/shows/reaction/interviews/till.html
Interview of Dr. Charles Till on PBS Frontline.

BN-600 LMFBR
  http://www.insc.anl.gov
  Select Reactor Maps/Maps of Nuclear Power Reactors/Russia/Beloyarskiy3.
A Roadmap for Developing Accelerator Transmutation of Waste (ATW) Technology:
  A Report to Congress
  http://www.pnl.gov/atw/
  reporttocongress/index2.html
  Download pdf file; DOE/RW-0519.
William Bebbington, "The Reprocessing of Nuclear Fuels," *Scientific American*,
  December 1976, p. 30. This classic popular article is relevant because there have
  been relatively few new technical developments.
Ted Breitmayer, "Taking Another Look at Reprocessing," *Radwaste Solutions*,
  September/October 2001, p. 49. A good review of the history and possible future
  of the subject.
Manson Benedict, Thomas H. Pigford, and Hans Wolfgang Levi, *Nuclear Chemical
  Engineering*, 2nd Ed., McGraw-Hill, New York, 1981.
A. J. Judd, *Fast Breeder Reactors: An Engineering Introduction*, Pergamon Press,
  Oxford, 1981.
Alan E. Waltar and Albert B. Reynolds, *Fast Breeder Reactors*, Pergamon Press, New
  York, 1981.

Chapter 15—Uranium Mill Tailings

Closing the Circle and *Linking Legacies*
  http://legacystory.apps.em.doe.gov
  Historical description of the U.S. weapons complex and environmental
  restoration.
A Citizen's Guide to Radon
  http://www.epa.gov/iaq/radon/pubs/citguide.html
  Subtitle: The Guide to Protecting Yourself and Your Family from Radon. Latest issue
  (2002) of EPA's information.
Radon Update
  http://www.physics.isu.edu/radinf/radon.htm#top
  Based on an article by Dr. A. B. Brill in *Journal of Nuclear Medicine*, February
  1994, provided by Physics Department, Idaho State University.
Radiation, Science, and Health
  http://cnts.wpi.edu/RSH/index.html
  Organization criticizes conservatism of standards advisory bodies and government
  regulators.
Merril Eisenbud and Thomas F. Gesell, *Environmental Radioactivity from Natural,
  Industrial, and Military Sources*, Academic Press, San Diego, CA, 1997.
Uranium: 1999 Resources, Production and Demand, Organisation for Economic
  Co-operation and Development, Paris, 2000. The so-called "Red Book," containing
  tables and analyses based on estimates made by each of 49 countries. Published
  biennially by OECD Nuclear Energy Agency and the International Atomic Energy
  Agency. Available for purchase and downloading (340 pp.) from the Web.
Jacob Shapiro, *Radiation Protection: A Guide for Scientists and Physicians*, 4th
  Ed., Harvard University Press, Cambridge, MA, 2002. Technical but not highly
  mathematical. Comprehensive yet practical. Includes radiation dose calculations
  for radium and radon and its daughters.
Bernard L. Cohen, *Radon: A Homeowner's Guide to Detection and Control*,
  Consumers Union, Mount Vernon, NY, 1987. By the publishers of *Consumer
  Reports*. Extensive tables of data on radon concentrations by state and county,
  and recommendations for reducing radon in homes.
National Research Council Staff, *Health Effects of Exposure to Radon* (BEIR VI),
  National Academy Press, Washington, D.C., 1998.

Chapter 16—Generation and Treatment of Low-Level Waste

Engineering and Design—Guidance for Low-Level Radioactive Waste (LLRW) and
Mixed Waste (MW) Treatment and Handling
http://www.usace.army.mil/inet/usace-docs/eng-manuals/em1110-1-4002/
Table of Contents of an online Engineer Manual of the U.S. Army Corps of
Engineers.

The Management of Low- and Intermediate-Level Radioactive Waste
http://www.nea.fr/html/brief/brief-06.html
One of NEA's Issue Briefs giving European perspective on nuclear issues. Topics
have numbers 01 to 10 in the URL, e.g., 08 is Probabilistic Safety Analysis.

*The Impact of Low-Level Radioactive Waste Management Policy on Biomedical
Research in the United States (2001)*
http://www.nap.edu/books/
0309073316/html/R1.html
National Academy Press online book. Review of a variety of methods of treatment
and discussion of recent developments.

Department of Energy Decommissioning Handbook
http://www.em.doe.gov/dd/decom_doc.html
Report DOE/EM-0383, January 2000. 137 pp. Available for download.

P. Andrew Karam, "Waste Management by a One-Man Band: Managing a University
and Medical LLRW Program," *Radwaste Solutions*, March/April 2000, p. 38.
Discusses practices and problems in a university research setting.

Melvin W. Carter, A. Alan Moghissi, and Bernd Kahn, *Management of Low-Level
Radioactive Waste*, Volumes 1 and 2, Pergamon Press, New York and Oxford,
1979. Comprehensive technical information that is still useful despite the
publication date.

Y. Bruce Katayama, Langdon K. Holton, Jr., Galen N. Buck, James F. Hutchens, and
Mark S. Culverhouse, "Advanced Remote Decontamination Techniques Reduce
Costs and Radiation Exposure," *Nuclear Technology*, July 1991. A clear
description of an actual decontamination. Advanced techniques and the results
are compared with those from earlier methods.

Jim McNeil, "Closing the Book: The Decommissioning of the Barnwell Nuclear Fuel
Plant," *Radwaste Solutions*, May/June 2000, p. 55. Gives the history of the Allied-
General Nuclear Services project and provides a thorough discussion of the
processes used to decontaminate.

Jim McNeil, "Plan Ahead, Establish Support Lines, and Be Prepared for Surprises,"
Radwaste Solutions, January/February 2001, p. 30. Emphasizes the virtue of
careful planning and coordination.

Russell A. Mellor, "Decommissioning: Thinking Through to the End," *Radwaste
Solutions*, September/October 2001, p. 26. Notes that with license renewal, the
amount of decommissioning of reactors is smaller than expected.

Chapter 17—Transportation of Radioactive Materials

Full-scale accident tests of casks
http://www.sandia.gov/tp/ SAFE_RAM/SEVERITY.HTM
Photos of tests of cask integrity in accident situations.

Regulatory Testing with Computer Analyses
http://www.sandia.gov/tp/SAFE_RAM/ANALYSIS.HTM
Links to theory, method, and comparison of designs with and without computer
simulation.

National Transportation Program
http://ntp.doe.gov
Department of Energy's web site. Access to program Transportation Routing
Analysis Geographic Information System (TRAGIS).

Analysis of transportation risk
    http://ttd.sandia.gov/nrc/modal.htm
    Select in Library/New Documents the reports *An Updated View of Spent Fuel Transportation Risk (Discussion Draft)* and *Spent Nuclear Fuel Transportation Package Performance Study.*
RADTRAN, Transportation Risk Assessment Computer Code
    http://ttd.sandia.gov/risk/radtran.htm
    Web access to the program via RADDOG.
Transportation of Spent Nuclear Fuel
    http://cnie.org/NLE/CRSreports/energy/eng-34.cfm
    Congressional Research Service Report by Mark Holt, 1998.
Transportation of Spent Nuclear Fuel and High-Level Waste to a Repository
    http://www.state.nv.us/nucwaste/trans/trfact03.htm
    Issues related to safety are raised by Nevada, which opposes Yucca Mountain.
Fact Sheet: Solutions for safe, efficient packaging and transportation
    http://nsnfp.inel.gov/fs-transportation.asp
    Links to several issues, including shipping cask integrity. From INEEL.
Package Performance Study References
    http://ttd.sandia.gov/nrc/docs.htm
    A large number of links to reports on transportation that can be downloaded. Includes material by detractors.
Nuclear materials transportation regulations
    http://www.nrc.gov/materials/transportation/regs-guides-comm.html
    Select Part 71.
Federal Radiological Emergency Response Plan
    http://www.au.af.mil/au/awc/awcgate/frerp/frerp.htm
    Operational plan as published in *Federal Register*, May 8, 1996, p. 20944, and June 5, 1996, p. 28583.
Emergency Preparedness and Response
    http://www.whitehouse.gov/deptofhomeland/sect4.html
    General scope of the Department of Homeland Security.
Final Environmental Statement on the Transportation of Radioactive Material by Air and Other Modes, NUREG-0170, Nuclear Regulatory Commission, 1977.
    Radiation hazards from transport of radioactive material are judged to be very small compared with those from normal transportation or from natural radiation background.
Marvin Resnikoff, *The Next Nuclear Gamble*, Council on Economic Priorities, New York, 1987. Subtitle: Transportation and Storage of Nuclear Waste. By a critic and opponent of nuclear power.

Chapter 18—Health, Safety, and Environmental Protection

Background Information Document Yucca Mountain Standards
    http://www.epa.gov/radiation/yucca/bid.htm
    History, background, and basis for regulation 40 CFR Part 197.
EPA Standards for Yucca Mountain
    http://www.epa.gov/radiation/yucca/regs.htm
    Select EPA's Public Health and Environmental Protection Standards for Yucca Mountain, Nevada. Downloads as pdf file.
United Nations Framework Convention and the Kyoto protocol
    http://www.unfccc.int
    Includes full texts of the documents.
    http://www.unfccc.int/resource/beginner.html
    Beginner's guide to the convention.
Earth Summit 2002 (Rio + 10)
    http://www.earthsummit2002.org
    Special Session of the General Assembly to Review and Appraise the Implementation of Agenda 21.

Raymond L. Murray, *Nuclear Energy*, 5th Ed., Butterworth-Heinemann, Boston, 2001. Elementary discussion of PRA, pp. 277-280.

David N. Savage, *The Scientific and Regulatory Basis for the Geological Disposal of Radioactive Wastes*, John Wiley & Sons, New York, 1996.

IAEA Staff, *Safety Assessment for Near-Surface Disposal of Radioactive Waste*, International Atomic Energy Agency, Vienna, 1999.

Systematic Approaches to Scenario Development, NEA/OECD, Paris, 1992. Identification of scenarios for performance assessment of waste disposal systems and discussion of modeling. Examples from United States, Canada, United Kingdom, and Sweden.

Confidence in Models of Radionuclide Transport for Site-specific Assessment, Proceedings of a 1999 workshop, NEA/OECD, Paris, 2001. A very recent collection of experiences in modeling around the world. Notes the need for greater use of natural analogues and for long-term experiments.

Richard E. Faw and J. Kenneth Shultis, *Radiological Assessment: Sources and Doses*, American Nuclear Society, La Grange Park, IL, 1999.

Samuel Glasstone and Walter H. Jordan, *Nuclear Power and Its Environmental Effects*, American Nuclear Society, La Grange Park, IL, 1980.

Geoffrey G. Eichholz, *Environmental Aspects of Nuclear Power*, Lewis Publishers, Chelsea, MI, 1985.

Dade W. Moeller, *Environmental Health, Revised Edition*, Harvard University Press, Cambridge, MA, 1997. Comprehensive coverage of all environmental topics, including radiation and energy production.

John E. Till and H. Robert Meyer, Eds., *Radiological Assessment, A Textbook on Environmental Dose Analysis*, Nuclear Regulatory Commission, Washington, D.C., NUREG/CR-3332, 1983. A comprehensive technical book describing techniques and models used to relate sources of radioactivity to effects on human health.

R. Allan Freeze and John A. Cherry, *Groundwater*, Prentice-Hall, Englewood Cliffs, NJ, 1979. A classic book on the science of water flow in geologic media and the transport of contamination.

H. C. Burkholder and E. L. J. Rosinger, "A Model for the Transport of Radionuclides and Their Decay Products Through Geologic Media," *Nuclear Technology*, Vol. 49, pp. 150-158, June 1980. An important paper, often cited.

A. G. Milnes, *Geology and Radwaste*, Academic Press, London, 1985. A comprehensive examination of the use of the earth in radioactive waste disposal. Expresses concern about the environmental impact of radioactive wastes.

William Miller, Russell Alexander, Neil Chapman, Ian McKinley, and John Smellie, *Natural Analogue Studies in the Geological Disposal of Radioactive Wastes*, Elsevier, Amsterdam, 1994. Reviews investigations at 14 sites around the world, including Oklo (natural fission reactor) and others (e.g., radioactive uranium and thorium ores and an ancient cache of iron nails).

Policy Implications of Greenhouse Warming, National Academy Press, Washington, D.C., 1991. A report by the Committee on Science, Engineering, and Public Policy. Provides information on the origin and nature of greenhouse gases and global warming. Considers adaptation and mitigation methods. Recommends continuing research.

Donald J. Wuebbles and Jae Edmonds, *Primer on Greenhouse Gases*, Lewis Publishers, Boca Raton, FL, 1991. Based on research at two institutions on the generation of gases, the relation to energy, and the effects on and of climate change. Extensive tables on the sources and properties of individual gases are presented.

Chapter 19—Disposal of Low-Level Wastes

Overview of Low-Level Waste
  http://www.nei.org
  Select Library/Fact Sheets/Disposal of Low-Level Radioactive Waste. By Nuclear Energy Institute.

Manifest Information Management System (MIMS)
   http://mims.mactec.com
   Tabulates data on shipments of LLW to disposal facilities up to 2001, when DOE
   discontinued reporting.
Chem-Nuclear Systems
   http://www.chemnuclear.com
   Description of disposal system.
Radioactive Waste Disposal Program (of Barnwell)
   http://www.state.sc.us/energy/llrwdisposal.htm
   Organization and administration of Barnwell, with many links. For costs, select
   rate schedule and download a pdf file.
National Profile of Commercially Generated Mixed Wastes
   http://www.epa.gov/radiation/mixed-waste/nat_prof.htm
   Describes report NUREG/CR-5938.
The Shallow Land Burial of Low-Level Radioactively Contaminated Solid Waste,
   Committee on Radioactive Waste Management, National Academy of Sciences,
   Washington, D.C., 1976. A document of historical interest.
Designs and Costs of Low-Level Waste Disposal Facilities, Rogers and Associates,
   Salt Lake City, UT, Report EPRI NP 53655, August 1987. Provides the
   classification system, conceptual designs, economics, and safety aspects.
J. Howard Kittel, Ed., *Near-Surface Land Disposal*, Harwood Academic Press, Chur,
   Switzerland, 1989. Technical information by specialists. Describes wastes, site
   selection, characterization, development, and operations. Options for greater-
   confinement disposal; environmental monitoring; corrective actions.
Edward R. Landa, "The First Nuclear Industry," *Scientific American*, November
   1982, p. 180. Description of the separation of radium for medical purposes.

Chapter 20—Disposal of Defense Wastes

Hanford Technical Library Electronic Reports
   http://www.pnl.gov/tech_lib/edocs.htm
   Links to reports including those about tanks.
National Transuranic Waste Program
   http://www.wipp.carlsbad.nm.us
   Information about the Waste Isolation Pilot Plant.
Performance Assessment of WIPP
   http://www.sandia.gov/E&E/risk/wast2.html
   Describes the tasks performed by Sandia National Laboratories.
EPA's WIPP Program
   http://www.epa.gov/radiation/wipp
   Regulatory aspects of the repository.
DOE Office of Environmental Management
   http://www.em.doe.gov
   Select Publications/List of Publications. Download pdf files *Baseline
   Environmental Management Report, 1996, Accelerated Cleanup: Pathways to
   Closure, From Cleanup to Stewardship*, and *Integrated Data Base Report*.
Closing the Circle and *Linking Legacies*
   http://legacystory.apps.em.doe.gov
   Historical description of the U.S. weapons complex and environmental
   restoration.
DOE's Office of Environmental Management information
   http://www.em.doe.gov/progint/info.html
   Select pdf document *Achieving Waste and Material Management and Site
   Cleanup More Efficiently.*
Vitrification of wastes at Savannah River Site
   http://www.srs.gov/general/srs-home.html
   Select News & Events/WSRNews& Events/FactSheets for information on saltstone
   and DWPF.

Megatons to Megawatts: A Snapshot
  http://www.usec.com/v2001_02/content/Megatons/RusGraphsnapshot.htm
  Flow diagram for U.S.-Russian program.
Radioactive Wastes at the Hanford Reservation, A Technical Review, National
  Academy of Sciences, Washington, D.C., 1978.
Wallace W. Schulz and Nicholas J. Lombardo, Eds., *Science and Technology for
  Disposal of Radioactive Tank Wastes*, Plenum Press, New York, 1998.
  Proceedings of a 1997 conference sponsored by DOE. Covers characterization,
  retrieval, pretreatment, immobilization, and process control and monitoring.
Kenneth A. Broberg and Robert Janke, "Groundwater Re-injection at Fernald,"
  Radwaste Solutions, September/October 2000, p. 19. Describes the system to
  decontaminate an aquifer.
Todd A. Nelson, "Dedicated to Cleanup: Environmental Remediation at Hanford,"
  Radwaste Solutions, July/August 2001, p. 17. Describes the transfer of
  contaminated material to the Environmental Restoration Disposal Facility, far
  from the Columbia River.
Kevin D. Crowley and John Ahearne, "Managing the Environmental Legacy of U.S.
  Nuclear-Weapons Production," *American Scientist*, November-December 2002,
  p. 514. Provides information on history of production, amounts of waste,
  locations, and plans for disposition. Notes that long-term stewardship will
  substitute for cleanup of radioactive sites.
Robert Noyes, *Nuclear Waste Cleanup Technology and Opportunities*, Noyes
  Publications, Park Ridge, NJ, 1995. Survey of national problem and available
  technologies, based on government reports.
Don J. Bradley, *Behind the Nuclear Curtain*, David R. Payson, Ed., Battelle Press,
  Columbus, OH, 1997. Subtitle: The Radioactive Waste Management in the Former
  Soviet Union. Insights on radioactive contamination resulting from neglect and
  discharges to the environment. Excellent diagrams, maps, tables, and photos.
Mark E. Byrnes, Editor, *Sampling and Surveying Radiological Environments*,
  Lewis Publishers, Boca Raton, FL, 2001. Equipment and methods that can be
  used successfully in decontamination and decommissioning of nuclear facilities.
  Emphasis is on EPA requirements.

Chapter 21—Disposal of Spent Fuel

Reference Design Description for a Geologic Repository, Rev. 01
  http://www.osti.gov/bridge
  Select Easy Search/Identifier Numbers, enter DOE/RW-98003756. Details on the
  site, handling operations, and barrier characteristics, as of 1997. The final system
  as described by later report revisions may differ somewhat.
Civilian Radioactive Waste Management Program Plan, Rev. 3 (2000)
  http://ocrwm.doe.gov
  Select Program Management, enter DOE/RW-0520, select pprev3.pdf. Includes
  objectives, goals, strategies, and planned activities, with schedule and costs.
  Appendices on history of the Civilian Radioactive Waste Management Program and
  excerpts from the Nuclear Waste Policy Act of 1982.
  Direct access http://www.ocrwm.doe.gov:80/pm/pdf/pprev3.pdf.
Yucca Mountain Resource Book
  http://www.nei.org
  Select above topic. From Nuclear Energy Institute.
Engineered Barrier System Materials Characterization
  http://ymp.llnl.gov/ebs.html
  Studies by Lawrence Livermore National Laboratory.
Viability Assessment of a Repository at Yucca Mountain, Nevada
  http://ocrwm.doe.gov:80/factsheets/doeymp0402.shtml
  Summary of findings as of December 1998 on site characteristics, design,
  performance, and costs. Need Adobe Acrobat Reader.

Yucca Mountain As a Radioactive Waste Repository
http://geopubs.wr.usgs.gov/circular/c1184
Circular 1184 of the U. S. Geological Survey, 1999. A thoughtful and informative discussion of the viability assessment for Yucca Mountain.
Recommendation by the Secretary of Energy Regarding the Suitability of the Yucca Mountain Site for a Repository Under the Nuclear Waste Policy Act of 1982. February 2002. Two sources:
  • *Federal Register*, Vol. 67, No. 39, February 27, 2002, 22 pp.
  • http://ocrwm.doe.gov:80/ymp/sr/sar.pdf.
Holly R. Trellue, "Does Transmutation of Spent Nuclear Fuel Produce More Hazardous Material Than It Destroys?" *Radwaste Solutions*, July/August 2002, p. 40. The answer is "No." Notes that a subcritical waste burner creates fission products. Paper provides graphs of toxicity of waste with time for various scenarios.
Micah D. Lowenthal, "Transmutation in the Nuclear Fuel Cycle: Approaches and Impacts," *Nuclear Technology*, June 2002, p. 284. Theory and calculations on the radiological hazard from wastes. Extensive list of references.
Hugh Benton and Judy Connell, "The Evolution of Spent-Fuel Waste Packages," *Radwaste Solutions*, March/ April 2001, p. 34. Describes the engineered barrier system designed to extend the life of the Yucca Mountain repository.
Geological Disposal of Radioactive Waste in Perspective, NEA/OECD, Paris, 2000. A monograph providing an overview of progress in legislation and regulation, scientific and technical programs, and a look at the future.
Geological Disposal of Radioactive Waste: Review of Developments in the Last Decade, NEA/OECD, Paris, 1999. Describes a questionnaire about advances and a digest of responses by 18 countries.
Technical Bases for Yucca Mountain Standards, National Academy Press, Washington, D.C., 1995. Statements on views on needed protection, performance assessment, and compliance. Appendix by Thomas Pigford.
James Flynn, et al., *One Hundred Centuries of Solitude: Redirecting America's High-Level Nuclear Waste Policy*, Westview Press, Boulder, CO, 1995. A critique of the nation's HLW program that recommends a moratorium on Yucca Mountain in favor of interim storage, voluntarily accepted by some community or state.
Kristin S. Schrader-Frechette, *Burying Uncertainty: Risk and the Case Against Geological Disposal of Nuclear Waste*, University of California Press, Berkeley, 1993. Key concepts discussed include value judgments, uncertainty, and equity. Supports a negotiated, monitored, retrieval storage (NMRS) approach with waiting for a century rather than immediate geological disposal.

Chapter 22—Laws, Regulations, and Programs

U.S. Federal Government Agencies Directory
http://www.lib.lsu.edu/gov/fedgov.html
A comprehensive set of links maintained by Louisiana State University libraries.
Federal Register Online Via GPO Process
http://www.access.gpo.gov/su_docs/aces/aces140.html
Database for 1995-2003.
Nuclear Regulatory Commission
http://www.nrc.gov
Explore drop-down menus. Select Electronic Reading Room/Document Collections/Regulations (10 CFR). Select 60 for geologic disposal, 61 for LLW, and 63 for Yucca Mountain.
Part 20-Standards for Protection against Radiation
http://www.nrc.gov/reading-rm/doc-collections/cfr/part020
Links to each subsection.

Environmental Protection Agency
  http://www.epa.gov
  Select About EPA/policy resources/ Policy Statements and Strategy Documents/Strategic Plan.
Status of Low-Level Radioactive Waste Disposal
  http://www.gao.gov
  Select GAO Reports, enter in Search RCED-99-238.
Low-Level Radioactive Waste Forum, Inc.
  http://www.llwforum.org
  A non-profit organization requiring membership or contribution.
Envirocare of Utah, Inc.
  http://www.envirocareutah.com
  Treatment and disposal of low-level waste, mixed wastes, and naturally occurring radioactive materials (NORM).
Risk-Informed Performance-Based Regulation
  http://www.nrc.gov
  Search on SECY-98-144 for an informative discussion of regulatory transition.
NRC Regulatory Guides
  http://www.nrc.gov
  Select Electronic Reading Room/Document Collections/Regulatory Guides (by Division)/Power Reactors.
  1.109 Calculation of Annual Doses to Man from Routine Releases of Reactor Effluents for the Purpose of Evaluating Compliance with 10CFR Part 50 Appendix I, October 1977.
  1.111 Methods for Estimation of Atmospheric Transport and Dispersion of Gaseous Effluents in Routine Releases from Light-Water-Cooled Reactors, July 1977.
  1.174 An Approach for Using Probabilistic Risk Assessment in Risk-Informed Decisions On Plant-Specific Changes to the Licensing Basis.
  1.175 An Approach for Plant-Specific, Risk-Informed Decisionmaking: Inservice Testing.
  Select Occupational Health.
  8.8 Information Relevant to Ensuring that Occupational Radiation Exposures at Nuclear Power Stations Will Be As Low As Reasonably Achievable, June 1978.
EPA's Yucca Mountain Standards
  http://www.epa.gov/docs/epacfr40/chapt-I.info/subch-F.htm
  Alternative: http://www.epa.gov
  Select Laws, Regulations & Dockets/Code of Federal Regulations/CFR Title 40: Protection of Environment/Chapter 1/Subchapter F/Part 197.
NRC's Yucca Mountain Review Plan
  http://www.nrc.gov/reading-rm/doc-collections/nuregs/staff/sr1804
  Report NUREG-1804, Rev. 2, describing the guidance to evaluate DOE's license application.
Transportation regulations on hazardous materials
  http://www.access.gpo.gov/nara/cfr/waisidx_01/49cfrv2_01.html
  DOT rules 49 CFR 171-179. Note same filename for each subpart.
U.S. Department of Energy
  http://www.energy.gov
  Select About Us or National Energy Policy.
U.S. Department of Energy laboratories
  http://phonebook.doe.gov/natlabs.html
  Links to home pages of all laboratories.
DOE's Yucca Mountain Suitability Guidelines
  http://www.access.gpo.gov/su_docs/aces/aces140.html
  Select 2001 Federal Register, Vol. 66/Final Rules and Regulations. Search term 10 CFR parts 960 and 963.

Nuclear Waste Technical Review Board
   http://www.nwtrb.gov
   Mission, membership, and reports.
DOE Energy Information Administration
   http://www.eia.doe.gov
   Example: For data, select By Fuel/Nuclear/U.S. Radioactive Waste and Spent
   Nuclear Fuel/Spent Nuclear Fuel.
International Atomic Energy Agency
   http://www.iaea.org/worldatom
   Choose one of the Programmes: Nuclear Applications; Nuclear Energy; Nuclear
   Safety & Security; Safeguards; and Technical Cooperation.
Nuclear Energy Institute
   http://www.nei.org
   Select About NEI/NEI Profile. Browse in Nuclear Technologies.
American Nuclear Society
   http://www.ans.org
   The organization and nuclear information. Select About or Publications.
   http://aboutnuclear.org
Nuclear Science and Technology and How It Influences Your Life.
   http://www.ans.org/pi/ps
   Position statements of the society.
Nuclear Energy Agency (OECD)
   http://www.nea.fr/welcome.html
   Press releases, new publications, new reports.
Nuclear Engineering Education Sourcebook
   http://www.ne.ncsu.edu/source/sourcebook03/index.html
   Includes links to lists of faculty in all institutions. John G. Gilligan, Ed.
Advisory Committee on Human Radiation Experiments-Final Report
   http://tis.eh.doe.gov/ohre/roadmap/achre/report.html
   Investigation by special committee.
Estimated exposures from iodine-131 in fallout
   http://rex.nci.nih.gov/massmedia/Fallout/index.html
   1997 National Cancer Institute report available for download.
   http://www.cdc.gov/nceh/radiation/fallout
   Links to progress report, feasibility study, and Q&A.
RadWaste.Org
   http://www.radwaste.org
   Links to a large number of Web sites. By Herne Data Systems, Inc.
Michael T. Ryan and Walter E. Newcomb, "The Status of Low-Level Radioactive
   Waste Management in the United States with Selected International Comparisons,"
   Technology 7: 479-489, 2000. (Web site of journal: http://www.nars.org/
   technology/index.html.) Paper gives an overview, with history of disposal facilities,
   information on DOE's LLW, regulatory requirements, and the stalled status and
   uncertain future for development efforts in the United States. Successes abroad
   are noted.
Alan Pasternak, "The Politics of Low-Level Radioactive Waste Disposal," *Radwaste
   Solutions*, May/June 2001, p. 16. A review of the national situation as of 2001.
   Notes that eventually only 14 states will have access to LLW disposal facilities.
Charles Judd, "Let the Private Sector Help," *Radwaste Solutions*, May/June 2000,
   p. 14. Discusses the situation on LLW disposal and tells how Envirocare can help.
Tye Rogers, "Careful Operation at Envirocare of Utah," *Radwaste Solutions*,
   May/June 2001, p. 11. Describes the way the privately owned facility meets all
   regulatory requirements.

Richard G. Telfer, "A New Role for the Nevada Test Site," *Radwaste Solutions*, March/April 2002, p. 48. Background on the use of the site for weapons tests and the new use by DOE for the disposal of LLW.

Michael E. Burns, Ed., *Low-Level Waste Regulation: Science, Politics and Fear*, Lewis Publishers, Chelsea, MI, 1988. Articles on all aspects. Considerably dated—expresses deep concern but optimism about LLW disposal.

Chapter 23—Societal Aspects of Radioactive Wastes

Society and Nuclear Energy: Towards a Better Understanding
http://www.nea.fr/html/ndd/reports/2002/nea3677.html
Select above title of French essay in New Reports.

U.S. Nuclear for the 21st Century
http://www.nei.org
Select Library/Speeches and Testimony/October 23, 2002 and June 10, 2002/ Joe Colvin. From Nuclear Energy Institute.

Annual Energy Outlook 2003 with Projections to 2025
http://www.eia.doe.gov/oiaf/aeo
Report DOE/EIA-0383 (2003) on all forms of energy.

Bernard L. Cohen, *The Nuclear Energy Option: An Alternative for the 90s*, Plenum Press, New York, 1990.

"Reversibility and Retrievability in Radioactive Waste Disposal," *Radwaste Solutions*, Part I, Nov./Dec. 2002, p. 39; Part II, Jan./Feb. 2003, p. 33. Concepts that may help achieve public confidence. Adapted from an OECD/NEA article.

Mary R. English, *Siting Low-Level Radioactive Waste Disposal Facilities: The Public Policy Dilemma*, Quorum Books, New York, 1992. A description of the processes in several states, with discussion of philosophic and practical issues. Key concepts are Authority, Trust, Risk, Justice, and Legitimacy. The author does not claim to have solutions but identifies the dimensions of the low-level waste problem.

Aaron Wildavsky, "No Risk Is the Highest Risk of All," *American Scientist*, January-February 1979, p. 32. An insightful article that asserts that too much caution results in less safety.

Richard G. Telfer, "Good Things Can Happen When the Public Gets Involved," *Radwaste Solutions*, July/August 2000, p. 45. Describes the roles of civic leaders and the public.

Chapter 24 Perspectives—A Summing Up

Civilian Nuclear Waste Disposal
http://www.cnie.org
One of the collection of Congressional Research Service reports made available by the National Library for the Environment. Select CRS/Waste Management/Report IB92059. Gives an overview of the political scene for all aspects of nuclear waste.

Index